ADVANCED FIELD THEORY

Micro, Macro, and Thermal Physics

ADVANCED FIELD THEORY

Micro, Macro, and Thermal Physics

HIROOMI UMEZAWA

Library of Congress Cataloging-in-Publication Data
Umezawa, H. (Hiroomi), 1924–
    Advanced field theory/Hiroomi Umezawa.
        p. cm.
    Includes bibliographical references and index.
    ISBN 978-1-56396-456-5
    1. Quantum field theory. 1. Title.
QC174.45U47   1992   92-40005
530.1'43-dc20   CIP

10   9   8   7   6   5   4   3   2   1

Contents

Preface

The fundamental interactions in physics at present are formulated in terms of the quantum field theory. This has led many physicists on an ambitious attempt at many kinds of unification mechanisms with a small number of fundamental axioms. Although success of these efforts is unpredictable, it has become more and more apparent that many different areas in physics seem to share numerous concepts which are intrinsically related to field theory. These include topics such as order-disorder, symmetry breakdown, correlation, collective modes, macroscopic quantum number, macroscopic quantum phenomena, etc. A remarkable aspect in this development is that the gap between micro and macro physics has narrowed. Therefore, it is worth trying to construct a cohesive picture over a wide area of physics on the basis of these concepts. This book is a modest attempt in this direction.

A unified perspective is not the sole purpose of this book. It may be useful also in other ways; for example, as a supplement to a regular course on quantum field theory. Though the existence of inequivalent Hilbert space representations, which sharply distinguish quantum field theory from quantum mechanics, is the basis of the physics of symmetry breakdown, ordered states and macroscopic quantum states, this subject is missing from regular text books and should be a part of a course on quantum field theory. If only for this purpose, the first four chapters should be taught. It is advisable to include also Chapter 5, which provides a brief description of the subject of broken symmetry and ordered states. I used an original draft of this book for such a course for PhD students in theoretical physics at the University of Alberta. According to this experience the first five chapters are suitable for a course of three and half months.

For those readers who are interested in macroscopic quantum phenomena the first six chapters give a good review. This book may be useful also for those interested in thermal quantum field theories, in which case they can skip Chapter 6. Readers interested only in equilibrium thermal quantum field theories may skip also Chapter 9.

Although the book is intended to provide a synthesis of physics, its particular attention is focused on thermo field dynamics, which is an operator formalism of thermal quantum field theories. Including recent advances, the presentation of thermo field dynamics is put in a systematic and coherent form.

The reference lists in this book are far from complete. It was very difficult to make reasonably complete lists of references because the subject of the book covers a wide scope in physics. However, the real reason for this is a fault on the author's side: I do not practice good bibliographic habits. I am sure that many important papers are missing from the lists. My sincere apology to those physicists whose articles are overlooked. I hope that readers would kindly inform me about missing articles. The references in Chapter 7 are mostly papers from my research group. This was not avoidable due to the reason that this chapter summarizes the early development of thermo field dynamics. However, the entire description in this chapter is reconstructed in a form smoothly connected to the main stream of the book which is based on the concept of the vacuum. Furthermore, I chose thermo field dynamics among many formulations of the thermal quantum theory, because it is also based on the same concept. The relationship between thermo field dynamics and other formulations of thermal quantum field theory is briefly summarized in Chapter 8. Since Chapter 9 treats new attempts in thermo field dynamics, there too the references are heavily sided to articles on thermo field dynamics.

During the period when this book was being prepared, the author was aided by many people, without whose help the writing of it would not have been possible. Dr. Yoshiya Yamanaka helped me constantly with valuable and useful advice. Dr. Timothy Evans made a thorough reading of the first draft of the book and Dr. Harry Schiff made a careful reading of the last draft of the book, providing comments and advice. Drs. Koichi Nakamura, Toshihiko Arimitsu and Henyou Chu also made valuable comments and useful suggestions. The book was considerably improved by the detailed and careful comments of all these people. A large part of Chapter 6 is based on an article prepared by Dr. Nicolas Papastamatiou and myself. My secretary, Mrs. Patricia Anderson, was also helpful in various aspects involved in the preparation of the manuscript. And lastly a special thanks to my editor, John Zumerchik, whose encouragement and guidance throughout were necessary to bring this work to completion. My heartful thanks go to all these people.

H. Umezawa
Edmonton
June 1992

1

Micro and Macro Physics

1.1 Micro, Macro and Thermal

Most of the phenomena that we observe in nature are of a macroscopic nature. A remarkable fact is the large variety of forms (or "phases") in which these macroscopic systems manifest themselves. Some simple examples are: gas → liquid → crystal with macroscopic defects; paramagnetic metals → ferromagnetic metals with domains, normal conductors → superconductors with or without vortex currents, etc.. Many of these phases are characterized by certain kinds of order such as the crystal order, magnetic order, superconducting order and so on. The definition of "order" needs a mathematical formulation of physics of symmetries and their breakdown which will be presented in chapter 5. For the moment we proceed with a vague picture by saying that a state which manifests a certain systematic pattern is called an ordered state. Such order is never perfect, because some defects always exist to violate the order.

This complex behavior of macroscopic systems raises a series of questions. Why can the same system appear in many kinds of phases? What is the mechanism for transitions between different phases? Why do ordered states accompany defects since the latter exhibit disorder in ordered states? These questions suggest that macroscopic systems may consist of a huge number of their fundamental elements and that many different phases correspond to the appearance of different states or aggregates of these fundamental elements. This fundamental world is called the microscopic world.

The physical picture for the microscopic world was originally a simple one. It was thought that every macroscopic system consists of a huge, but still a finite number of small particles called atoms. It is then a challenging task to derive the macroscopic properties from the physics for the microscopic world and thus to find answers to all of the questions mentioned above. However, this turns out to be a complex task because the microscopic world is not governed by the laws of classical physics which describes the macroscopic world. We now know that these macroscopic systems consist of "fundamental entities" whose dynamics are ruled by the laws of quantum physics. Our immediate question, then, is to ask how microscopic systems ruled by the laws of quantum physics can manifest macroscopic phenomena ruled by the laws of classical physics:

micro   →   macro
(quantum physics)   (classical physics)

Once we understand this mechanism, we would have a very powerful method for the analysis of materials in condensed matter physics. Furthermore, we might be able to appreciate still more fascinating phenomena. One is the understanding of the remarkably sophisticated and stable orders observed in living systems. In particular, intelligent use of memories in the human brain suggests a very sophisticated mechanism for controlling orders. Another direction is to cultivate an overall picture of the development and the different orders of the Universe since its birth.

A particularly remarkable set of properties common to all macroscopic systems are known as thermal effects. The study of thermal properties in classical physics led to the beautiful classical physics called thermodynamics. It is well known that thermal effects play decisive roles in controlling phase transitions between the different ordered states. Classical thermodynamics did provide a phenomenological description of phase transitions. It may be reasonable to hope that clarification of the micro-macro mechanism may disclose the microscopic origin of thermal properties. This is a goal of statistical mechanics. Statistical mechanics was based on the belief that macroscopic objects are seen only through macroscopic observation in which one only observes some kind of statistical average of the very complicated behavior of huge number of microscopic objects; the microscopic behavior is so complicated that no observation is able to follow the behavior of each microscopic particle. It was hoped that this averaging mechanism would explain the thermal properties. Behind this view was the old atomism based on the picture of huge but finite number of microscopic particles.

The term "quantum physics" might suggest quantum *mechanics*. Indeed, quantum mechanics was the first form of quantum theory historically. Quantum mechanics inherited from classical atomism the picture of a finite number of particles or a finite number of degrees of freedom. However, even in classical physics, the concept of a finite number of particles did not cover the entire domain; the concept of a " field" with an infinite number of degrees of freedom prevailed in electromagnetism which is incorporated into classical field theory. The quantum version of field theory is quantum field theory. Quantum physics includes both quantum mechanics and quantum field theory. To clarify the above mentioned relation between the microscopic and macroscopic world, we need the modern approach to both classical physics and quantum physics.

1.2   Classical Physics for Particles and Fields

Consider an isolated system of particles in classical physics. Let n denote the number of particles. The motion of each particle is described by the temporal behavior of its position and momentum vectors. We denote the position vectors and momentum vectors of the n particles by $\vec{x}_i$ and $\vec{p}_i$ with $i = 1, 2, \cdots, n$. Let us assume that the particles move in a potential V, which is a function of the n position vectors, from which the forces $\vec{f}_i$ acting on the particles are obtained as

$$\vec{f}_i = -\vec{\nabla}_i V. \tag{1.1}$$

The dynamics of this system is controlled by the equations

$$\dot{\vec{p}}_i = \vec{f}_i, \tag{1.2}$$
$$m_i \dot{\vec{x}}_i = \vec{p}_i, \tag{1.3}$$

where dot means the time derivative. Thus the motion of n particles is described by n trajectories in the $6n$ dimensional space called the phase space for $2n$ vectors $(\vec{x}_i, \vec{p}_i)$. These trajectories are determined when the $2n$ vectors at any given time, say t_0, are specified. In the Hamiltonian formalism these equations, (1.2) and (1.3), are obtained from the Lagrangian $\mathcal{L} = \sum (1/2) m_i \dot{x}_i^2 - V$ by means of the variational method. Since $\partial \mathcal{L} / \partial \dot{\vec{x}}_i = m_i \dot{\vec{x}}_i$, which is equal to $\vec{p}_i$ according to (1.3), $\vec{p}_i$ is the canonical conjugate of $\vec{x}_i$.

The Hamiltonian $H = \sum \dot{\vec{x}}_i \vec{p}_i - \mathcal{L}$ is obtained from this Lagrangian as

$$H = \sum_i \frac{1}{2m_i} p_i^2 + V, \tag{1.4}$$

which is the energy of the system. The above equations, (1.2) and (1.3), lead to the energy conservation law $\dot{H} = 0$. This is associated with the invariance of these equations under time translation $t \rightarrow t + c$ with a constant c. Time translational invariance reflects the fact that the choice of origin of the time coordinate for an isolated system is arbitrary. Since the choice of origin of the spatial coordinates is also arbitrary, the equations, (1.2) and (1.3), are (and therefore the Lagrangian is) invariant under the spatial translation $\vec{x}_i \rightarrow \vec{x}_i + \vec{c}$ with any constant vector $\vec{c}$. This means that the potential is a function of mutual differences of position vectors as $V = V(\vec{x}_1 - \vec{x}_2, \vec{x}_1 - \vec{x}_3, \cdots)$. We then have $\sum \vec{\nabla}_i V = 0$, which is the law of total momentum conservation

$$\sum_i \dot{\vec{p}}_i = 0. \tag{1.5}$$

Furthermore, the arbitrariness of choice of the direction of spatial coordinate for an isolated system leads to invariance of the Lagrangian under any rotation of the spatial coordinates, and in its turn this leads to the conservation of the total angular momentum $\sum \dot{\vec{l}_i} = 0$ with the angular momenta $\vec{l}_i = \vec{x}_i \times \vec{p}_i$. In this way the invariant properties of the Lagrangian gives rise to conservation laws. This universal relationship between invariance and conservation law is inherited by quantum physics and plays a decisive role in developments throughout this book.

Exercise 1 Making use of the Hamiltonian-Jacobi formalism, derive the above conservation laws by assuming only the invariance of the Lagrangian under time and space translations and under spatial rotations without assuming the above particular form of the Lagrangian.

The development of special relativity showed that the above form of classical equations for particles in the so called Newtonian mechanics, is true only for objects with small velocities. The correct theory including high velocity objects is special relativity. When we consider free particles, the Hamiltonian is given by

$$H = \sum_i \sqrt{p_i^2 + m_i^2}. \tag{1.6}$$

Here we use units such that $c = 1$. The combination $(H - \sum_i m_i)$ with this form of H reproduces the Hamiltonian of Newtonian mechanics for small $\vec{p}_i$.

So far we have considered a system of particles. However the particle picture does not cover the entire domain of classical physics. Electromagnetic radiation propagates as a wave rather than behaving as a particle. This wave character is well accommodated by the concept of a field, and the most common formalism for electromagnetic phenomena is the Maxwell formalism of electromagnetic fields. *In field theory the state of a system of fields at any given time is specified by measurements of a complete set of observables over the entire space at that time.* Thus the possible states of a system of fields are described by a basic set of functions of the space coordinate $\vec{x}$. These basic functions depend also on time and their temporal behavior describes the time development of the state of the fields. In electromagnetism the basic functions are the electric field $\vec{E}(\vec{x}, t)$ and the magnetic field $\vec{H}(\vec{x}, t)$. These functions are controlled by the Maxwell equations:

$$\vec{\nabla} \times \vec{E} + \frac{1}{c}\frac{\partial \vec{H}}{\partial t} = 0, \tag{1.7}$$

$$\vec{\nabla} \times \vec{H} - \frac{1}{c}\frac{\partial \vec{E}}{\partial t} = \frac{1}{c}\vec{j} \tag{1.8}$$

$$\vec{\nabla} \cdot \vec{E} = \rho, \tag{1.9}$$

$$\vec{\nabla} \cdot \vec{H} = 0. \tag{1.10}$$

Here $\vec{j}$ and ρ are the electric current and electric charge density, respectively.

To simplify the above equations, let us introduce the notations:

$$\partial^i = \frac{\partial}{\partial x_i}, \quad \partial^0 = \frac{1}{c}\frac{\partial}{\partial t}. \tag{1.11}$$

and

$$\partial_\mu = g_{\mu\nu}\partial^\nu. \tag{1.12}$$

Here the suffix μ runs over $(0, 1, 2, 3)$ and $g_{\mu\nu}$ is the diagonal matrix with the diagonal elements $g_{00} = -1, g_{11} = g_{22} = g_{33} = 1$. In the following a repeated suffix implies summation over the suffix, unless otherwise stated. For example,

$$\partial^2 \equiv \partial^\mu \partial_\mu, \tag{1.13}$$

$$= \vec{\nabla}^2 - \frac{1}{c^2}\frac{\partial^2}{\partial t^2}. \tag{1.14}$$

In the following, units where $c = 1$ will be frequently used.

Now the Maxwell equations read

$$\partial^\mu F_{\mu\nu} = j_\mu, \tag{1.15}$$

$$\epsilon^{\mu\nu\sigma\rho}\partial_\nu F_{\sigma\rho} = 0. \tag{1.16}$$

Here $F_{\mu\nu}$ is defined by $E_i = F_{0i}$ and $F_{ij} = \epsilon^{ijk}H_k$ with the condition $F_{\mu\nu} = -F_{\nu\mu}$. The ϵ^{ijk} is antisymmetric with respect to exchange of any two suffices and is fixed by $\epsilon^{123} = 1$. Thus, $F_{12} = H_3, F_{23} = H_1, F_{31} = H_2$. Since $F_{\mu\nu}$ forms a 4×4 antisymmetric matrix, it has six independent elements; three of them give $\vec{H}$ and the others form $\vec{E}$. The $\epsilon^{\mu\nu\sigma\rho}$ is defined by the condition that it changes its sign under exchange of any two suffices together with the condition $\epsilon^{0123} = 1$. The suffix such as μ can be brought up by the metric tensor $g^{\mu\nu}$ as $F_\mu{}^\sigma = g^{\sigma\nu}F_{\mu\nu}$. The four dimensional current j_μ is defined by $(\vec{j}, c\rho)$. As it is known that each suffix such as μ acts as the suffix of a four dimensional vector under the Lorentz transformation, the above expression of the Maxwell equation is of a covariant form in the Minkowsky geometry. In other words, the Maxwell theory has relativistic invariance. This led to the development of the special relativity.

It is known that the electric and magnetic fields can be expressed in terms of the four dimensional vector potential $A_\mu(\vec{x}, t)$ as

$$F_{\mu\nu} = \partial_\mu A_\nu - \partial_\nu A_\mu. \tag{1.17}$$

The quantities $F_{\mu\nu}$, and therefore the Maxwell equations, are, invariant under the gauge transformation $A_\mu \to A_\mu + \partial_\mu \lambda$ with a function $\lambda(\vec{x}, t)$ *which is required to have a well defined Fourier transform.* This condition is important because *it guarantees that the gauge function satisfies the condition*:

$$[\partial_\mu, \partial_\nu]\lambda = 0, \tag{1.18}$$

which is the reason for the gauge invariance of $F_{\mu\nu}$. As an example of a function which violates this condition (1.18), consider a cylindrical angle θ around the third axis. It gives

$$\vec{\nabla} \times \vec{\nabla}\theta = \vec{e}_3\delta(x_1)\delta(x_2). \tag{1.19}$$

This is due to the fact that this angle is multivalued with a topological singularity along the third axis. The above condition of the existence of a Fourier transform excludes these topological functions. It also excludes functions which diverge in the infinitely large $\vec{x}$ region, because Fourier transformable functions are square integrable

$$\int d^3x |\lambda|^2 < \infty. \tag{1.20}$$

To study electromagnetic radiation, we assume $j_\mu = 0$. We can then choose the particular gauge such that $A_0 = 0$. Then the Maxwell equations become

$$\partial^2 \vec{A} = 0, \tag{1.21}$$

$$\vec{\nabla} \cdot \vec{A} = 0. \tag{1.22}$$

The relation (1.21) shows that the $\vec{A}$ propagates as a wave and (1.22) shows that it is a transverse wave with two components. Therefore, the vector potential has the following Fourier form:

$$\vec{A}(x) = \frac{1}{(2\pi)^{3/2}} \int d^3k \, \frac{1}{\sqrt{2\omega_k}}[\vec{a}_k e^{i(\vec{k}\cdot\vec{x}-\omega_k t)} + \vec{a}_k^\dagger e^{-i(\vec{k}\cdot\vec{x}-\omega_k t)}], \tag{1.23}$$

where x is $(\vec{x}, t)$ and $\omega_k = |\vec{k}|$. The dagger means the complex conjugate. The suffix k for oscillator amplitudes means $\vec{k}$. Here $\vec{a}_k$ has two components $(a_{i,k}; i = 1, 2)$ orthogonal to the direction of $\vec{k}$.

The Lagrangian which gives rise to the Maxwell equations is

$$\mathcal{L} = -\int d^4x \frac{1}{4}F^{\mu\nu}F_{\mu\nu} \tag{1.24}$$

$$= \int d^4x \frac{1}{2}[E^2 - H^2]. \tag{1.25}$$

With the present choice of gauge $A_0 = 0$, this Lagrangian shows that the canonical conjugate of $\vec{A}(x)$ is $\vec{E}$. This leads to the Hamiltonian

$$H = \int d^3x \frac{1}{2}[\vec{E}^2 + \vec{H}^2]. \tag{1.26}$$

This is the energy of the radiation field. Feeding the Fourier form (1.23) into this, the Hamiltonian becomes

$$H = \sum_{i=1,2} \int d^3k \frac{1}{2}\omega_k [a_{i,k}a_{i,k}^\dagger + a_{i,k}^\dagger a_{i,k}]. \tag{1.27}$$

Here, we carefully preserved the order among a and $a^\dagger$ in preparation for quantization. It is obvious that $a_{i,k}\exp[-i\omega_k t]$ is an oscillating function.

To clarify a relation between the particle picture and wave picture, let us change the oscillator variables $(a_{i,k}, a_{i,k}^\dagger)$ to the real variables $(q_{i,k}, p_{i,k})$ as

$$q_{i,k} = \frac{1}{\sqrt{2\omega_k}}\left(a_{i,k}e^{-i\omega_k t} + a_{i,k}e^{i\omega_k t}\right), \tag{1.28}$$

$$p_{i,k} = \frac{\sqrt{\omega_k}}{\sqrt{2}i}\left(a_{i,k}e^{-i\omega_k t} - a_{i,k}e^{i\omega_k t}\right). \tag{1.29}$$

We then have

$$\dot{q}_{i,k} = p_{i,k}, \tag{1.30}$$
$$\dot{p}_{i,k} = -\omega_k^2 q_{i,k}. \tag{1.31}$$

Comparing this with the equations (1.1), (1.2) and (1.3), we see that the radiation field Fourier amplitude with each $(i, \vec{k})$ acts as a one dimensional particle with unit mass and with the potential $V_{i,k} = (1/2)\omega_k^2 q_{i,k}^2$ and $p_{i,k}$ is the canonical momentum conjugate to $q_{i,k}$. Indeed, the Hamiltonian in (1.27) is

$$H = \sum_{i=1,2} \int d^3k \frac{1}{2}[p_{i,k}^2 + \omega_k^2 q_{i,k}^2], \tag{1.32}$$

which is the aggregate of the particle Hamiltonian with unit mass and the oscillator potential $V_{i,k}$. We state that the radiation field is mathematically equivalent to an assembly of infinite numbers of simple Harmonic oscillators.

This *does not mean* that the field is equivalent to an assembly of particles, because $q_{i,k}$ and $p_{i,k}$ are endowed with no concept of particle position and momentum, but are the degrees of freedom associated with the amplitudes $a_{i,k}$ and $a_{i,k}^\dagger$. In quantum mechanics the position vector is replaced by the q-number position q, and therefore, *no c-number position variable is left*. Contrary to this, the field $\vec{A}$ depends on the c-number position coordinate $\vec{x}$ which is of vital significance in describing the wave propagation in the space. *This spatial position $\vec{x}$ has nothing to do with the above $q_{i,k}$.* This shows that the above similarity between a particle and field is only a mathematical analogy; the particle picture and field picture have a sharp distinction. We will soon see a significant implication of this fact. However, this mathematical similarity helps us to apply the same quantization method to both particles and fields.

Before we proceed to the quantization, a note about the electromagnetic theory may be worth mentioning here. Since we treated in the above only the pure radiation field (i.e. $j_\mu = 0$), we were able to choose the gauge conditioned by $A_0 = 0$. When we preserve the variable A_0, the Lagrangian density $(1/4)F^{\mu\nu}F_{\mu\nu}$ indicates that A_0 does not have its canonical conjugate. This causes a complication in quantization of the electromagnetic field, because the quantization is based on the canonical formalism. One way to obtain a canonical formalism for the electromagnetic field is to introduce an auxiliary field B and add the term $\mathcal{L} = B\partial^\mu A_\mu - (1/2)\alpha B^2$ to the Lagrangian [1, 2, 3, 4]. The new Lagrangian makes B to be the canonical momentum conjugate of A_0, reviving the canonical formalism on basis of which the quantization is performed. The gauge fields together with quark fields constitute the most significant part of the fundamental world in the present day physics and electromagnetism is only a simple part of this world. The above mentioned complication in the quantization of electromagnetic fields also occurs with these gauge fields in general. Since a study of gauge fields is not a major target of this book and since there are many excellent books (see, for example, [5]) on the subject of gauge fields, we are not going to pursue this subject any further.

1.3   Quantum Mechanics and Quantum Field Theory

According to a common historical view quantum mechanics and quantum field theory evolved through the two steps of quantization: the first and second quantization. However, from a physics viewpoint, there was only one set of quantization rules in this development; the same quantization rules were applied to two different objects, that is, particles and fields. With these quantization rules one assumes the commutation relation

$$[q,p] = i\hbar, \tag{1.33}$$

for a set of canonical coordinates q and their momenta p. Here, $\hbar$ is the Planck constant divided by 2π. With this commutation relation, a canonical set of classical c-numbers (i.e. ordinary numbers) (q,p) are replaced by q-numbers (quantum operators). Thus the dynamical quantities consisting of q and p such as the Hamiltonian, angular momentum and others become also q-numbers. We then apply the theory of linear algebras and their representations to these q-numbers, and the states of systems are represented by the state vectors in the Hilbert space. The quantization rules dictate also that the complex conjugate (i.e. dagger conjugate) is to be understood as the Hermitian conjugate.

When these quantization rules are applied to a system of N particles discussed in the last subsection, we have

$$[q_i^a, p_j^b] = i\hbar\delta_{ij}\delta_{ab}, \tag{1.34}$$

where q_i^a means the a-th ($a = 1, 2, 3$) component of the i-th ($i = 1, 2, \cdots, n$) particle position vector. The p_j^b can be understood accordingly. Replacing $\vec{x}_i$ by $\vec{q}_i$ in the Hamiltonian in (1.4), the Hamiltonian H becomes a q-number consisting of $\vec{q}_i$ and $\vec{p}_i$. The angular momentum is also the q-number $\sum \vec{q}_i \times \vec{p}_i$. Thus the c-number position variables $\vec{x}_i$ completely disappear from quantum mechanics. In classical mechanics the temporal behavior of $\vec{p}_i$ and $\vec{x}_i$ (which are now replaced by $\vec{q}_i$) was controlled by the equations (1.2) and (1.3). With the help of the commutation relation (1.34), we can rewrite these equations as

$$\dot{\vec{q}_i} = [\vec{q}_i, H], \tag{1.35}$$

$$i\dot{\vec{p}_i} = [\vec{p}_i, H]. \tag{1.36}$$

Since any operator, say $\mathcal{O}$, in this quantum mechanics consists of $\vec{q}_i$ and $\vec{p}_i$, its time dependence is controlled by the equation

$$i\dot{\mathcal{O}} = [\mathcal{O}, H]. \tag{1.37}$$

Exercise 2 Show that the equations (1.35) and (1.36) give (1.2) and (1.3).

In case of the radiation field the quantization rule was applied, not to the position vector, but to the canonical variables $(q_{i,k}, p_{i,k})$ which consists of the radiation amplitudes $(a_{i,k}, a_{i,k}^\dagger)$ according to (1.28) and (1.29). Considering the fact that $\vec{k}$ is a vector consisting of three continuous numbers, we write

$$[q_{i,k}, p_{j,l}] = i\hbar\delta_{ij}\delta^3(\vec{k} - \vec{l}). \tag{1.38}$$

Here (i, j) signifies the two polarizations of radiation. The use of the Dirac δ-function implies that both sides of this equation should be smeared out with a square-integrable test function (i.e. wave packet). As will be shown later, this use of wave packet is required by the concept of probability which dictates that the state vectors should have a well defined norm. With help of the relations (1.28) and (1.29) the commutation relation (1.38) leads to the following commutation relations for the oscillator operators $(a_{i,k}, a_{i,k}^\dagger)$:

$$[a_{i,k}, a_{j,l}^\dagger] = \delta_{ij}\delta(\vec{k} - \vec{l}). \tag{1.39}$$

Now the vector potential $\vec{A}$ is not a c-number but a q-number with the Fourier form (1.23) in which the amplitudes $(a_{i,k}, a_{i,k}^\dagger)$ are the operators satisfying the oscillator commutation relation (1.39). Note that we have the operator $\vec{A}(\vec{x})$ still a function of c-number $\vec{x}$. The Hamiltonian is given by (1.27), which is now a q-number. With use of the commutation relation in (1.39), we can put equation (1.21) in the following form:

$$i\dot{\vec{A}} = [\vec{A}, H], \tag{1.40}$$

$$i\ddot{\vec{A}} = [\dot{\vec{A}}, H]. \tag{1.41}$$

This can be extended to any operator as it was in (1.37):

$$i\dot{\mathcal{O}} = [\mathcal{O}, H], \tag{1.42}$$

where $\mathcal{O}$ is a function of $\vec{A}$.

Through the above consideration we see a remarkable difference between quantum mechanics and quantum field theory. When quantum mechanics was created, the c-number position vector $\vec{x}$ and its momentum $\vec{k}$ are replaced by the operators $\vec{q}$ and $\vec{p}$ respectively, leaving us without any c-number spatial variable. Furthermore, in quantum mechanics we do not have any c-number variable associated with a given time, therefore there is no hope of creating spatially inhomogeneous classical objects in the framework of quantum mechanics. On the other hand in the case of field theory, quantization is applied to the amplitudes, while the position vector $\vec{x}$ remains as a c-number; $\vec{A}$ is an operator and a function of c-number $\vec{x}$. This is an essential reason why many classical objects are created in a system of quantum fields, but not in quantum mechanical systems.

Furthermore, in quantum mechanics the number of sets of canonical operators is the same as the number of particles and, therefore, is finite, while a quantum field carries an infinite number of degrees of freedom. This is another fact which makes the physics of quantum mechanics sharply different from that of quantum field theory. In later chapters we will learn that the presence of infinite number of degrees of freedom in quantum field theory is the origin of many different macroscopic quantum states, which narrows the gap between the microscopic and macroscopic physics.

To simplify the situation, in the following we use scalar fields $\varphi(x)$ to illustrate our points. We also use units where $\hbar = c = 1$, unless otherwise stated. When this is a free field, its Fourier form is

$$\varphi(x) = \frac{1}{(2\pi)^{3/2}} \int d^3k \, \frac{1}{\sqrt{2\omega_k}} [a_k e^{i(\vec{k}\cdot\vec{x} - \omega_k t)} + b_k^\dagger e^{-i(\vec{k}\cdot\vec{x} - \omega_k t)}], \tag{1.43}$$

with the oscillator operators a_k and b_k satisfying

$$[a_k, a_l^\dagger] = \delta(\vec{k} - \vec{l}), \tag{1.44}$$

$$[b_k, b_l^\dagger] = \delta(\vec{k} - \vec{l}), \tag{1.45}$$

$$[a_k, b_l^\dagger] = 0. \tag{1.46}$$

The quantum particle energy ω_k is not specified. To generalize our consideration to include spin freedom and both fermion and boson statistics is straightforward.

We now present a brief summary of field quantization. First note the equal-time commutation relation

$$[\varphi(x), \Pi_\varphi(y)]\delta(t_x - t_y) = i\delta^{(4)}(x - y), \qquad (1.47)$$

where $\Pi_\varphi = \dot{\varphi}^\dagger$.

Exercise 3 Prove (1.47) by means of the Fourier form (1.43) together with the oscillator commutation relation (1.44).

The free field equation for φ is

$$[\partial_t^2 + \omega^2(-i\vec{\nabla})]\varphi = 0, \qquad (1.48)$$

where $\omega(-i\vec{\nabla})$ is defined by

$$\omega(-i\vec{\nabla})e^{i\vec{k}\cdot\vec{x}} = \omega_k e^{i\vec{k}\cdot\vec{x}}. \qquad (1.49)$$

The Lagrangian is

$$\mathcal{L} = \int d^4x [\dot{\varphi}^\dagger \dot{\varphi} - \varphi^\dagger \omega(-i\vec{\nabla})\varphi]. \qquad (1.50)$$

This Lagrangian shows that $\Pi_\varphi = \dot{\varphi}^\dagger$ is the canonical momentum field of φ, because $\Pi_\varphi = \partial\mathcal{L}/\partial\dot{\varphi}$. Thus, (1.47) shows that the canonical quantization rules can be applied, not only to the oscillator operators, but also to the field operator. In the relativistic case we have $\omega_k = \sqrt{k^2 + m^2}$ with a mass parameter m.

The ideas above can be extended to interacting fields which will be denoted by ψ. Start from a classical field Lagrangian and identify the canonical momentum field. Then, assume the canonical commutation relation of the form (1.47). Derive the Hamiltonian H also from the Lagrangian. The time derivative of any operator then satisfies the relation in (1.42). This is the method used by Heisenberg and Pauli in their first paper on the quantum electrodynamics in 1929 [6, 7]. The Fourier form (1.43) is extended to the interacting cases in the form

$$\psi(x) = \frac{1}{(2\pi)^{3/2}} \int d^3k \frac{1}{\sqrt{2\omega_k}}[a_k(t)e^{i\vec{k}\cdot\vec{x}} + b_k^\dagger(t)e^{-i\vec{k}\cdot\vec{x}}]. \qquad (1.51)$$

An interaction effect makes the time dependence of $a_k(t)$ more complicated than the simple oscillator form $\exp[-i\omega_k t]$, but the oscillator commutation relation for equal-time is still required:

$$[a_k(t), a_l^\dagger(t)] = \delta(\vec{k} - \vec{l}), \qquad (1.52)$$

$$[b_k(t), b_l^\dagger(t)] = \delta(\vec{k} - \vec{l}), \qquad (1.53)$$

$$[a_k(t), b_l^\dagger(t)] = 0. \qquad (1.54)$$

The Fourier form of the canonical momentum of ψ is

$$\Pi_\psi(x) = -i\frac{1}{(2\pi)^{3/2}} \int d^3k \frac{\sqrt{\omega_k}}{\sqrt{2}} [b_k(t)e^{i\vec{k}\cdot\vec{x}} - a_k^\dagger(t)e^{-i\vec{k}\cdot\vec{x}}]. \qquad (1.55)$$

The ψ and Π_ψ satisfy the equal time canonical commutation relation of the form (1.47) due to (1.52).

Fields with the Fourier form (1.51) are called scalar fields of type 2. The type 1 scalar fields have the Fourier form

$$\psi(x) = \frac{1}{(2\pi)^{3/2}} \int d^3k \, a_k(t)e^{i\vec{k}\cdot\vec{x}} \qquad (1.56)$$

with $a_k(t)$ satisfying the equal time commutation relation (1.52).

Quantum fields with the commutation relation (1.52) are boson fields. In the case of fermion fields (1.52) is replaced by the anticommutation relations:

$$[a_k(t), a_l^\dagger(t)]_+ = \delta(\vec{k} - \vec{l}), \qquad (1.57)$$

where use was made of the notation

$$[A, B]_+ = AB + BA. \qquad (1.58)$$

There are many other kinds of fields such as the Dirac field, the superconducting electron fields, the Duffin - Kemmer fields, etc.. For a general description of fields, see for example [4, 8]. In this book we do not touch the description of general kinds of fields. This book is written in such a manner that a detailed knowledge of these fields is not required.

1.4   The c-q Transmutation Condition

It was pointed out in the last subsection that the c-number position variable $\vec{x}$ survived through quantization as a c-number in quantum field theory. At first this would seem to contradict the fundamental requirement in quantum theory that any change is induced by a transformation in a Hilbert space, say $\mathcal{H}$. Since $u^{-1}cu = c$ for any c-number c and any transformation operator u, c-numbers do not change under any transformation. However, more thought indicates that this is not a contradiction but is an indication of a more elaborate requirement. This requirement dictates the following: any c-number change should be created by a q-number change, i.e., by a transformation in $\mathcal{H}$. This requirement is called the c-q transmutation condition [9, 4]. Then an immediate question is to ask what transformation in $\mathcal{H}$ induces a change in the c-number variable $\vec{x}$ (i.e. spatial translation). The answer can be seen from the combination $a_k \exp[i\vec{k} \cdot \vec{x}]$ in the Fourier form of the scalar field in (1.43). Consider the operator

$$\vec{P} = \int d^3k \, \vec{k}a_k^\dagger a_k. \qquad (1.59)$$

Using the general formula

$$e^{-A}Be^{A} = B + [B, A] + \frac{1}{2!}[[B, A], A] + \cdots, \qquad (1.60)$$

we obtain from the equation (1.44)

$$e^{-i\vec{c}\vec{P}}a_k e^{i\vec{c}\vec{P}} = a_k e^{i\vec{k}\cdot\vec{c}}, \qquad (1.61)$$

where $\vec{c}$ is a c-number vector.

Exercise 4 Derive the relation in (1.60).

Hint: Replace A by θA with a c-number continuous parameter θ and study the derivatives with respect to θ.

This shows that the operator $\vec{P}$ generates a change in the operator $\vec{a}_k$ *which in its turn induces the spatial translation $\vec{x} \rightarrow \vec{x} + \vec{c}$.* This is the c-q transmutation for the spatial translation. In this way the c-number spatial variables survive in quantum field theory without violating the c-q transmutation condition and open a way to create classical objects.

In quantum theory any operator which generates spatial translation is called a momentum operator. Since $\vec{P}$ generates the spatial translation, it is called the total momentum operator.

A similar consideration applied to time-translations shows that the generator is

$$H_0 = \int d^3k\, \omega_k a_k^\dagger a_k. \qquad (1.62)$$

This is the Hamiltonian for the free scalar field. As was pointed out in the last section, this Hamiltonian is the same as the one obtained from the Lagrangian for the free scalar field.

We see that the c-q transmutation condition plays a decisive role in the survival of the c-number position variable and prepares for the creation of macroscopic objects in quantum field theory. We will see in later sections other examples in which the c-q transmutation condition plays a significant role in the control of the behavior of c-number variables.

1.5 Matter and the Vacuum

In the development of the theory of electromagnetic fields, it was once assumed that the vacuum was full of a material called ether. This assumption was motivated by a belief that wave propagation needs a medium. This medium was called the ether. With many assumptions about the properties of the ether, the vacuum became rich in matter. However, the concept of the rest system of the ether was not compatible with the Michelson - Morley experiment and was eliminated in the course of development of special relativity.

However, the development of quantum field theory revived the idea of rich vacuum. According to quantum field theory, a large number of particles may condense in vacua, giving rise to space-time dependent macroscopic objects. For example, a condensation of particles with spin-up creates a macroscopic magnetic moment in up-direction and the vacuum exhibits an object of ferromagnetic type.

This new development reminds us of the old story of the ether and leads us to ask the question; how can the presence of the rest system of the particle condensate be consistent with the principle of relativity. The answer to this question lies in the sophisticated mechanism of so called spontaneously broken symmetries which is controlled by the celebrated Nambu-Goldstone theorem. When the space-time dependent macroscopic form of the condensate spontaneously breaks the Lorentz symmetry, there appear certain Goldstone zero-energy modes which act as an agent for maintaining the Lorentz invariance. To our surprise these zero-energy modes appear to be quantum mechanical modes. In this way, the condensation mechanism in quantum field theory creates not only classical objects but also quantum mechanical modes. We will explain this in later chapters.

1.6 REFERENCES

[1] N. Nakanishi, *Prog. Theor. Phys.* **49** (1973) 640.

[2] N. Nakanishi, *Prog. Theor. Phys.* **50** (1973) 1388.

[3] B. Lautrup, *Mat. Fys. Medd. Dan. Vid. Salsk.* **35** (1967) 1.

[4] H. Umezawa, H. Matsumoto, and M. Tachiki, *Thermo Field Dynamics and Condensed States*, (North-Holland, Amsterdam, 1982).

[5] N. Nakanishi and I. Ojima, *Covariant Formalism of Gauge Theories and Quantum Gravity*, (World Scientific, Singapore, 1990).

[6] W. Heisenberg and W. Pauli, *Zeits. f. Phys.* **56** (1929) 1.

[7] W. Heisenberg and W. Pauli, *Zeits. f. Phys.* **59** (1929) 168.

[8] H. Umezawa, *Quantum Field Theory*, (North-Holland, Amsterdam, 1956).

[9] H. Matsumoto, N. J. Papastamatiou, G. Semenoff, and H. Umezawa, *Phys. Rev.* **D24** (1981) 406.

2

Vacuum Correlation

2.1 Number Representation

To understand some of the profound differences between quantum mechanics and quantum field theory, we sketch, in this section, the quantum mechanics for a finite number of canonical variables. Let us start from the relation between canonical operators and oscillator operators mentioned in section 1.2.

When we have an oscillator operator a satisfying

$$[a, a^\dagger] = 1, \qquad (2.1)$$

the operators

$$q = \frac{1}{\sqrt{2}}(a + a^\dagger), \qquad (2.2)$$

$$p = \frac{1}{\sqrt{2}i}(a - a^\dagger) \qquad (2.3)$$

satisfy the canonical commutation relation

$$[q, p] = i. \qquad (2.4)$$

Thus a system with a canonical variable acts like an oscillator.

Define the number operator

$$N = a^\dagger a \qquad (2.5)$$

and denote its orthonormalized eigenvectors by $|n\rangle\rangle$:

$$N|n\rangle\rangle = n|n\rangle\rangle, \quad \langle\langle n|m\rangle\rangle = \delta_{nm}. \qquad (2.6)$$

Exercise 1 Considering the fact that $a^\dagger$ is the Hermitian conjugate of a, prove that n is non-negative.

Exercise 2 Prove that $a|n\rangle\rangle$ is an eigenvector of N with eigenvalue $n - 1$ and that $a^\dagger|n\rangle\rangle$ is one with eigenvalue $n + 1$.

Hint: make use of the oscillator commutation relation (2.1)

Since n cannot be negative, we see that n has a minimum value. Denote by $|0\rangle\rangle$ the eigenvector of N with the minimum eigenvalue. Then we should have

$$a|0\rangle\rangle = 0, \qquad (2.7)$$

because action of a decreases n. This gives rise to

$$N|0\rangle\rangle = 0, \tag{2.8}$$

which implies that the minimum eigenvalue of N is zero. This justifies the notation $|0\rangle\rangle$ which is $|n\rangle\rangle$ with $n = 0$. The state $|0\rangle\rangle$ is called the vacuum.

Then, m-times repeated action of $a^\dagger$ on $|0\rangle\rangle$ creates a state with an integer $n = m$. This proves that n assumes all the non-negative integers. It is due to this reason that n is frequently interpreted as the particle number. The name *vacuum* for $|0\rangle\rangle$ is due to the fact that its particle number is zero. Since $a^\dagger$ increases n by one, it is called the creation operator. The a decreases n by 1, acting as the annihilation operator. The state $|n\rangle\rangle$ is created by n operations of $a^\dagger$ on the vacuum state $|0\rangle\rangle$:

$$|n\rangle\rangle = \frac{1}{\sqrt{n!}}(a^\dagger)^n|0\rangle\rangle, \tag{2.9}$$

Here, the coefficient is determined by the normalization condition (the second equation in (2.6)).

The state vector space $\mathcal{H}$ is formed by all the normalizable superposition of the state vectors $\{|n\rangle\rangle\}$; $\mathcal{H} = \{\sum c_n|n\rangle\rangle : \sum |c_n|^2 < \infty\}$. This space is called the Fock space. Thus, we say that the set $\{|n\rangle\rangle\}$ is the basis of the Fock space $\mathcal{H}$. Since n assumes all the non-negative integers, the representation with this base is also called the number representation.

Introduce the notation

$$\Delta A = A - \langle A \rangle \tag{2.10}$$

with the definition

$$\langle A \rangle = \langle i|A|i \rangle, \tag{2.11}$$

where $|i\rangle$ stands for any state vector in $\mathcal{H}$. Then, the uncertainty principle states that

$$\langle (\Delta p)^2 \rangle \langle (\Delta q)^2 \rangle \geq (1/4). \tag{2.12}$$

Note that we use the unit $\hbar = 1$.

2.2   Coherent, Squeezed or Thermal-Like States

2.2.1   COHERENT STATES

Consider a single oscillator with associated operator a. We introduced the vacuum $|0\rangle\rangle$ in section 2.1. However, the problem in choosing a vacuum is complicated by the fact that *there can be many different kinds of vacuum*.

To understand this problem, let us recall the quantum electrodynamics in section 1.3. There it was shown that the radiation field is an assembly of photons. However, in phenomena in quantum electrodynamics we find not only the states of photons only, but also a variety of forms of macroscopic classical radiations. Our question now is to ask how one can obtain macroscopic radiation waves from quantum electrodynamics. One may intuitively suspect that these macroscopic radiations may be formed by condensation of many photons. This suggests that states of these macroscopic objects are not the eigenstates of particle numbers, but states with large uncertainty in particle numbers.

This reminds us of the problem of phase and number which has attracted attention of many physicists in the past. This problem arose from the anticipation that the phase and photon number might be canonical conjugate to each other. Let us begin with a crude consideration which leads us to such an anticipation. Suppose the wave function of a single photon and denote it by $\varphi(x)$. Then the wave function of n-photons is the product $\varphi(x_1) \cdots \varphi(x_n)$. This wave function changes its phase factor by $\exp[in\theta]$ when each photon wave function acquires the phase factor $\exp[i\theta]$. Thus, the operation of $-i\partial/\partial\theta$ on the n-photon wave function just multiplies the wave function by the photon number n. This suggests that $n \ (= -i\partial/\partial\theta$ in a θ representation) might be the canonical momentum conjugate to θ. Another similar argument is given by the formula

$$e^{-i\theta N} a e^{i\theta N} = e^{i\theta} a \tag{2.13}$$

with the number operator

$$N = a^\dagger a. \tag{2.14}$$

This shows that N generates the phase change, suggesting that the number would be the canonical conjugate of the phase. To understand this, recall that the momentum is the generator of spatial shift. If this anticipation would be justified, it would give the uncertainty relation

$$\Delta\theta\Delta n \geq \frac{\hbar}{2}, \tag{2.15}$$

which would mean that when the phase is specified, the photon number is completely uncertain; to create a radiation wave with a specific phase, such as a laser wave, we would need to have many photons. Unfortunately it is difficult to formulate this argument in a mathematical form, because the phase has not been defined as an operator. This is not a decisive objection to the above phase-number uncertainty relation, when we recall the fact that in the usual quantum theory we have no time operator, although we have the time-energy uncertainty relation. The construction of phase operator is a subject studied by many physicists. See, for examples, [1, 2, 3, 4, 5, 6]. The reason why it is difficult to define the phase or time operator is the fact that the particle number or energy spectrum is non-negative and, therefore, is lower bounded. This situation is drastically different from the one in case of the position and momentum whose eigenvalue spectra are lower unbounded. In the last section of this chapter we explain some aspect of this subject by relying mainly on the article of Ban in [6].

Here we study a state representing a macroscopic radiation which contains many photons through the boson condensation mechanism. Let us denote this state by $|0\rangle$. It is obvious that this is not the vacuum $|0\rangle\rangle$ of a-operator, because $a|0\rangle$ cannot vanish. However, we may look for the operator α, which satisfies $\alpha|0\rangle = 0$ and $[\alpha, \alpha^\dagger] = 1$. If we found such α, the state $|0\rangle$ would be the vacuum associated with α oscillator, although it contains many a-quanta. In the following we show an example of such α. Thus, the vacuum appropriate for the wave with a specific phase, such as a laser outputs, is a new vacuum in which many photons are condensed. This example shows how particle condensation can modify the vacuum. In this way there exist many kinds of vacua.

A well known form of vacuum with particle condensation is the coherent state, which is the subject of this subsection.

We denote the vacuum associated with a by $|0\rangle\rangle$, as we did in section 2.1:

$$a|0\rangle\rangle = 0. \tag{2.16}$$

We now perform the transformation

$$a \rightarrow \alpha(\theta) = a + \theta. \tag{2.17}$$

with a c-number θ. This is called the boson translation or the Bogoliubov transformation for coherent states. The vacuum for $\alpha(\theta)$ is denoted by $|0(\theta)\rangle$:

$$\alpha(\theta)|0(\theta)\rangle = 0. \tag{2.18}$$

We then have

$$a|0(\theta)\rangle = -\theta|0(\theta)\rangle. \tag{2.19}$$

The state $|0(\theta)\rangle$ is called the coherent state of a with the eigenvalue $-\theta$ [7].

We have

$$\alpha(\theta) = U_c(\theta)aU_c^{-1}(\theta)) \tag{2.20}$$

with

$$U_c(\theta) = \exp[iG_c(\theta)]. \tag{2.21}$$

in which the generator $G_c(\theta)$ is given by

$$G_c(\theta) = -i(\theta^* a - \theta a^\dagger). \tag{2.22}$$

Indeed, use of the formula (1.60) shows that (2.20) gives the boson translation (2.17). This gives

$$|0(\theta)\rangle = U_c(\theta)|0\rangle\rangle. \tag{2.23}$$

We can rewrite $U_c(\theta)$ as

$$U_c(\theta) = \exp\left(-\frac{1}{2}|\theta|^2\right)\exp\left(-\theta a^\dagger\right)\exp\left(\theta^* a\right). \tag{2.24}$$

Exercise 3 Derive (2.24).

Hint: prove

$$\frac{\partial}{\partial \rho}\left[U_c(\theta)e^{-\theta^* a}\right] = -U_c(\theta)e^{-\theta^* a}\frac{\partial}{\partial \rho}[\theta a^\dagger + \frac{1}{2}\rho^2], \qquad (2.25)$$

where ρ is $|\theta|$.

Now (2.23) and (2.24) give

$$|0(\theta)\rangle = \exp\left(-\frac{1}{2}|\theta|^2\right)\exp\left(-\theta a^\dagger\right)|0\rangle. \qquad (2.26)$$

This indicates that the α-vacuum state, $|0(\theta)\rangle$, is a superposition of states with many a-particles. This is an example of *condensation* of a-particles. In the same way as we built the a-Fock space (denoted by $\mathcal{H}(a)$) by operations of powers of $a^\dagger$ on $|0\rangle$, we can build the Fock space $\mathcal{H}(\theta)$ by cyclic operation of $\alpha^\dagger(\theta)$ on $|0(\theta)\rangle$. We see that $\mathcal{H}(a)$ *is equivalent to* $\mathcal{H}(\theta)$ in the sense that any vector in $\mathcal{H}(a)$ is a superposition of vectors in $\mathcal{H}(\theta)$ and vice versa. We have the parameterized Fock space $\mathcal{H}(\theta)$ such that any two spaces with different θ are equivalent to each other. *The coherent state saturates the uncertainty relation:*

$$\langle(\Delta p)^2\rangle\langle(\Delta q)^2\rangle = (1/4). \qquad (2.27)$$

2.2.2 SQUEEZED STATES

There are many reasons for need of a radiation wave packet of very small size, because such a wave packet makes the position measurement very precise. One such application is in the observation of the tiny effect of a gravitational wave. When a wave packet has the property either of

$$\langle(\Delta p)^2\rangle \leq \frac{1}{2} \qquad (2.28)$$

or of

$$\langle(\Delta q)^2\rangle \leq \frac{1}{2}, \qquad (2.29)$$

the wave is called a squeezed wave. In this subsection we study the structure of the vacuum state for squeezed waves.

To do this we consider bilinear generators consisting of a^2 and $a^{\dagger 2}$:

$$G_s = i\frac{1}{2}[a^2 - a^{\dagger 2}]. \qquad (2.30)$$

Defining

$$U_s(\theta) = \exp[i\theta G_s] \qquad (2.31)$$

with a real parameter θ, we make the transformation

$$\alpha(\theta) \quad = \quad U_s(\theta)aU_s^{-1}(\theta), \tag{2.32}$$
$$\alpha^\dagger(\theta) \quad = \quad U_s(\theta)a^\dagger U_s^{-1}(\theta). \tag{2.33}$$

which reads as

$$\alpha(\theta) \quad = \quad ca - da^\dagger, \tag{2.34}$$
$$\alpha^\dagger(\theta) \quad = \quad ca^\dagger - da. \tag{2.35}$$

Here the real parameter c and d are given by $c = \cosh\theta$ and $d = \sinh\theta$ respectively. This transformation is called the Bogoliubov transformation for the squeezed state. The commutation relations do not change under this transformation.

The vacuum of $\alpha(\theta)$ is denoted by $|0(\theta)\rangle$: $\alpha(\theta)|0(\theta)\rangle = 0$. We then have $|0(\theta)\rangle = U_s(\theta)|0\rangle$, which gives

$$|0(\theta)\rangle = \exp\left[-\frac{1}{2}\ln\cosh\theta\right]\exp\left[\frac{1}{2}a^{\dagger 2}\tanh\theta\right]|0\rangle. \tag{2.36}$$

Exercise 4 Derive (2.36).

Hint: Define

$$f_n = \langle\langle 0|a^{2n}U_s(\theta)|0\rangle\rangle, \tag{2.37}$$

and derive the differential equation

$$\frac{\partial}{\partial\theta}f_n = -\frac{1}{2}f_{n+1} + n(2n-1)f_{n-1}, \tag{2.38}$$

which has the solution

$$f_n = C_n e^{-(1/2)\ln\cosh\theta}(\tanh\theta)^n \tag{2.39}$$

with $C_n = C_0[(2n-1)!/(n-1)!](1/2^n)$.

The relation (2.36) indicates that the α-vacuum state, $|0(\theta)\rangle$, is a super-position of states with even number of a-particles. This is another example of the condensation of a-particles. We again see that $\mathcal{H}(a)$ is equivalent to $\mathcal{H}(\theta)$ in the sense that any vector in $\mathcal{H}(a)$ is a superposition of vectors in $\mathcal{H}(\theta)$ and vice versa. Here $\mathcal{H}(\theta)$ is the Fock space created by operations of power of $\alpha(\theta)^\dagger$ on $|0(\theta)\rangle$. We again have the parameterized Fock space $\mathcal{H}(\theta)$ such that any two spaces with different θ are equivalent to each other.

Introducing the canonical operators (q, p) through (2.2) and (2.3), we calculate their uncertainty by using the relations (2.10) and (2.11) with the i-state being the squeezed state (i.e. α-vacuum). Then, we obtain

$$\langle(\Delta q)^2\rangle \quad = \quad (1/2)(c+d)^2, \tag{2.40}$$
$$\langle(\Delta p)^2\rangle \quad = \quad (1/2)(c-d)^2. \tag{2.41}$$

Exercise 5 Derive these relations (2.40) and (2.41). Hint: Make use of the relations (2.34) and (2.35)

The above result shows that *the squeezed state saturates the uncertainty relation*:

$$\langle(\Delta p)^2\rangle\langle(\Delta q)^2\rangle = 1/4. \tag{2.42}$$

Since c and d increase with θ up to infinity, we can make $\langle(\Delta p)^2\rangle$ as small as we wish by choosing positive and large θ. We can make $\langle(\Delta q)^2\rangle$ very small by using negative θ. The name "squeezed" state originates from this fact.

2.2.3 TWO MODE SQUEEZED STATES

In the last subsection we studied a one mode squeezed state. In this subsection two mode squeezed states are analyzed.

Let us now consider a generator consisting of two commuting sets of oscillators, a and $\tilde{a}$:

$$[a, a^\dagger] = [\tilde{a}, \tilde{a}^\dagger] = 1, \tag{2.43}$$

$$[a, \tilde{a}] = [a, \tilde{a}^\dagger] = 0. \tag{2.44}$$

We introduce the generator

$$G_B = i[a\tilde{a} - \tilde{a}^\dagger a^\dagger]. \tag{2.45}$$

and define $U_B(\theta) = \exp[i\theta G_B]$. Then, the transformation

$$\alpha(\theta) = U_B(\theta)a U_B^{-1}(\theta), \tag{2.46}$$

$$\tilde{\alpha}(\theta) = U_B(\theta)\tilde{a} U_B^{-1}(\theta), \tag{2.47}$$

with the real parameter θ gives

$$\alpha(\theta) = ca - d\tilde{a}^\dagger, \tag{2.48}$$

$$\tilde{\alpha}(\theta) = c\tilde{a} - da^\dagger, \tag{2.49}$$

with c-number parameters c and d being given by $c = \cosh\theta$ and $d = \sinh\theta$, respectively. This is called the Bogoliubov transformation for two mode squeezed states [8, 9, 10, 11, 12, 13]. The commutation relations do not change under this transformation. Note that

$$c^2 - d^2 = 1. \tag{2.50}$$

The relations (2.48) and (2.49) can be put in the form

$$\begin{bmatrix} a \\ \tilde{a}^\dagger \end{bmatrix}^\mu = B^{-1}(\theta)^{\mu\nu} \begin{bmatrix} \alpha(\theta) \\ \tilde{\alpha}^\dagger(\theta) \end{bmatrix}^\nu, \tag{2.51}$$

$$[a^\dagger, -\tilde{a}]^\mu = [\alpha^\dagger(\theta), -\tilde{\alpha}(\theta)]^\nu B(\theta)^{\nu\mu} \tag{2.52}$$

with the Bogoliubov matrix $B(\theta)$ of the form

$$B(\theta) = \begin{bmatrix} c & -d \\ -d & c \end{bmatrix}. \tag{2.53}$$

It is easy to prove the relation

$$B^{-1}(\theta) = \tau_3 B(\theta) \tau_3, \tag{2.54}$$

where use was made of the Pauli matrices $\tau_i; i = 1, 2, 3$. Thus

$$\tau_3 = \begin{bmatrix} 1 & 0 \\ 0 & -1 \end{bmatrix}. \tag{2.55}$$

It is convenient to make use of the following doublet notation,

$$
\begin{aligned}
a^{\mu} : a^1 &= a, \; a^2 = \tilde{a}^{\dagger}, & (2.56) \\
\bar{a}^{\mu} : \bar{a}^1 &= a^{\dagger}, \; \bar{a}^2 = -\tilde{a}, & (2.57) \\
\alpha(\theta)^{\mu} : \alpha(\theta)^1 &= \alpha(\theta), \; \alpha(\theta)^2 = \tilde{\alpha}^{\dagger}(\theta), & (2.58) \\
\bar{\alpha}(\theta)^{\mu} : \bar{\alpha}(\theta)^1 &= \alpha^{\dagger}(\theta), \; \bar{\alpha}(\theta)^2 = -\tilde{\alpha}(\theta), & (2.59)
\end{aligned}
$$

because then the above transformation reads as

$$
\begin{aligned}
a^{\mu} &= B^{-1}(\theta)^{\mu\nu} \alpha(\theta)^{\nu}, & (2.60) \\
\bar{a}^{\mu} &= \bar{\alpha}(\theta)^{\nu} B(\theta)^{\nu\mu}. & (2.61)
\end{aligned}
$$

The vacuum of $(a, \tilde{a})$ is denoted by $|0\rangle\rangle$, while the one for $(\alpha(\theta), \tilde{\alpha}(\theta))$ by $|0(\theta)\rangle\rangle$;

$$
\begin{aligned}
a|0\rangle\rangle &= \tilde{a}|0\rangle\rangle = 0, & (2.62) \\
\alpha(\theta)|0(\theta)\rangle\rangle &= \tilde{\alpha}(\theta)|0(\theta)\rangle\rangle = 0. & (2.63)
\end{aligned}
$$

Then, we have $|0(\theta)\rangle\rangle = U_B(\theta)|0\rangle\rangle$, which gives

$$|0(\theta)\rangle\rangle = \exp\left[-\ln\cosh\theta\right] \exp\left[a^{\dagger}\tilde{a}^{\dagger}\tanh\theta\right]|0\rangle\rangle. \tag{2.64}$$

Exercise 6 Derive (2.64).
   Hint: Define

$$f_n = \langle\langle 0|[a\tilde{a}]^n U_B(\theta)|0\rangle\rangle. \tag{2.65}$$

Then derive the differential equation

$$\frac{\partial}{\partial\theta} f_n = -f_{n+1} + n^2 f_{n-1}, \tag{2.66}$$

whose solution is

$$f_n = C_n e^{-\ln\cosh\theta}(\tanh\theta)^n \tag{2.67}$$

with $C_n = n!C_0$.

The relation (2.64) indicates that *in $|0(\theta)\rangle$ the $(a\tilde{a})$-pairs are condensed.* The state $|0(\theta)\rangle$ is called the two mode squeezed state of $(a, \tilde{a})$.

We can build the $(a, \tilde{a})$-Fock space (denoted by $\mathcal{H}(a, \tilde{a})$) by operations of powers of $a^{\dagger}$ and $\tilde{a}^{\dagger}$ on $|0\rangle\rangle$, while the Fock space $\mathcal{H}(\theta)$ is given by operations of $\alpha^{\dagger}(\theta)$ and $\tilde{\alpha}^{\dagger}(\theta)$ on $|0(\theta)\rangle$. We see that $\mathcal{H}(a, \tilde{a})$ is equivalent to $\mathcal{H}(\theta)$ and therefore that the parameterized Fock spaces $\mathcal{H}(\theta)$ forms an equivalent set.

Since the $\alpha(\theta)$-vacua are the states with a condensation of $a\tilde{a}$-pairs, they are invariant under exchange of a and $\tilde{a}$. This exchange is called tilde conjugation. Thus we state that the vacua are invariant under tilde conjugation:

$$\langle 0(\theta)|^{\sim} = \langle 0(\theta)|, \quad |0(\theta)\rangle^{\sim} = |0(\theta)\rangle. \tag{2.68}$$

We introduce the number parameters by $n \equiv \langle 0(\theta)|a^{\dagger}a|0(\theta)\rangle$ and $\tilde{n} \equiv \langle 0(\theta)|\tilde{a}^{\dagger}\tilde{a}|0(\theta)\rangle$. We then have

$$n = \tilde{n} = d^{2}. \tag{2.69}$$

This expresses the parameter $d = \sinh\theta$ in terms of the number parameter n. The matrix B is therefore characterized by the number parameter.

Exercise 7 Derive (2.69).
Hint: Make use of the relations (2.48) and (2.49).

Analysis of the uncertainty relations discloses a very elaborate mechanism through which a correlation between the two oscillators is induced. We will see that the two mode squeezed states exhibits one particular aspect of the possible quantum vacuum effects, one which is closely related to the thermal effect. It is due to this reason that the two mode squeezed Bogoliubov transformation is called also the thermal Bogoliubov transformation. This plays a decisive role in construction of a thermal quantum field theory called thermo field dynamics (TFD). We therefore continue to discuss the two mode squeezed transformation in the following sections.

2.2.4 AN EQUIVALENCE THEOREM

When we consider all possible generators which are no more than quadratic in oscillator operators, all the transformations which modify the vacuum are built from products of the three kinds of transformations used above; the two mode squeezed (or the thermal) transformation, the one mode squeezed transformation and the coherent state transformation. They provide us with a basic technique for dealing with general problems involving the so called thermal coherent squeezed states in quantum optics.

We close this section with a comment of vital significance. The examples in this section illustrate the general statement that all the state vector spaces for normalizable states for a set of a *finite* number of canonical operators are equivalent. That is, any vector in one space can be expressed in terms of a well defined sum of vectors of another vector space. This implies that *we have practically only one Fock space, because all the Fock spaces are equivalent.*

2.3   Noise in Pure States

2.3.1   THERMAL-LIKE NOISE IN TWO-MODE STATE

In this section it will be shown that the bosonic Bogoliubov transformation, (2.48) and (2.49), creates a noise in a pure state [13].

To do this we now move from the oscillator variables to the canonical variables by using (2.2) and (2.3):

$$q \;=\; \tfrac{1}{\sqrt{2}}(a+a^\dagger), \; p \;=\; \frac{1}{\sqrt{2i}}(a-a^\dagger), \tag{2.70}$$

$$\tilde{q} \;=\; \tfrac{1}{\sqrt{2}}(\tilde{a}+\tilde{a}^\dagger), \; \tilde{p} \;=\; -\frac{1}{\sqrt{2i}}(\tilde{a}-\tilde{a}^\dagger), \tag{2.71}$$

and

$$q(\theta) \;=\; \tfrac{1}{\sqrt{2}}\left(\alpha(\theta)+\alpha^\dagger(\theta)\right), \; p(\theta) \;=\; \frac{1}{\sqrt{2i}}\left(\alpha(\theta)-\alpha^\dagger(\theta)\right), \tag{2.72}$$

$$\tilde{q}(\theta) \;=\; \tfrac{1}{\sqrt{2}}\left(\tilde{\alpha}(\theta)+\tilde{\alpha}^\dagger(\theta)\right), \; \tilde{p}(\theta) \;=\; -\frac{1}{\sqrt{2i}}\left(\tilde{\alpha}(\theta)-\tilde{\alpha}^\dagger(\theta)\right). \tag{2.73}$$

Thus, $(q,p,\tilde{q},\tilde{p})$ is $(q(\theta),p(\theta),\tilde{q}(\theta),\tilde{p}(\theta))$ with $\theta = 0$. Recall tilde conjugation which was defined in the last section; tilde conjugation exchanges a and $\tilde{a}$. Here we use tilde conjugation in a generalized form; the relations for $\tilde{A}$ are obtained from those for A through the tilde conjugation rules, which exchange a with $\tilde{a}$ and replaces the imaginary unit i with $-i$. This change of a c-number by its complex conjugate is the new rule added to the list of tilde conjugation rules. The new form of the tilde conjugation rule is convenient because it preserves the form of Bogoliubov transformation, as we will soon find out. From the above definitions, one can prove the canonical commutation relations:

$$[q,p] \quad = -[\tilde{q},\tilde{p}] \quad = i \tag{2.74}$$

$$[q(\theta),p(\theta)] \;=\; -[\tilde{q}(\theta),\tilde{p}(\theta)] \;=\; i. \tag{2.75}$$

Here too the tilde conjugation rule for c-numbers is respected. However, these relations show that the tilde momentum is not $\tilde{p}$ but $(-\tilde{p})$. With the simple linear relations obtained from (2.51), (2.52), (2.70), (2.71), (2.72) and (2.73) we again find the Bogoliubov transformation

$$\begin{bmatrix} q \\ \tilde{q} \end{bmatrix}^\mu = B^{-1}(\theta)^{\mu\nu} \begin{bmatrix} q(\theta) \\ \tilde{q}(\theta) \end{bmatrix}^\nu, \; \begin{bmatrix} p \\ \tilde{p} \end{bmatrix}^\mu = B^{-1}(\theta)^{\mu\nu} \begin{bmatrix} p(\theta) \\ \tilde{p}(\theta) \end{bmatrix}^\nu. \tag{2.76}$$

Note that, if the tilde conjugation rule for c-numbers were not respected in the above definitions of canonical operators, this Bogoliubov transformation of canonical operators would not be realized.

Exercise 8 Prove (2.76).

Hint: use the Bogoliubov transformation (2.51) and (2.52) for oscillator operators and the definitions of the canonical operators in terms of the oscillator operators

With the doublet notations $(q^1 = q, q^2 = \tilde{q}), (p^1 = p, p^2 = \tilde{p}), (q^1(\theta) = q(\theta), q^2(\theta) = \tilde{q}(\theta))$ and $(p^1(\theta) = p, p^2(\theta) = \tilde{p}(\theta))$, (2.76) reads as

$$q^\mu = B^{-1}(\theta)^{\mu\nu} q(\theta)^\nu, \quad p^\mu = B^{-1}(\theta)^{\mu\nu} p(\theta)^\nu. \tag{2.77}$$

On the other hand, (2.54) holding for (2.53) implies

$$B^{-1}(\theta) \tau_3 B^{-1}(\theta) = \tau_3. \tag{2.78}$$

Therefore the following quantity is invariant under the Bogoliubov transformation:

$$I \equiv \langle q^\mu q^\nu \rangle \tau_{3\mu\sigma} \tau_{3\nu\rho} \langle p^\sigma p^\rho \rangle \tag{2.79}$$

with the notation $\langle A \rangle = \langle i|A|i \rangle$ where $|i\rangle$ can be any state. Since the Bogoliubov transformation (2.77) is linear, we have the same transformation as the above for $(\Delta q^\mu, \Delta p^\mu, \Delta q(\theta)^\mu, \Delta p(\theta)^\mu)$, where we used the notation $\Delta A = A - \langle A \rangle$. Replacing q etc. by Δq etc. in I in (2.79), we obtain

$$\langle \Delta q^\mu \Delta q^\nu \rangle \tau_{3\mu\sigma} \tau_{3\nu\rho} \langle \Delta p^\sigma \Delta p^\rho \rangle = \langle \Delta q(\theta)^\mu \Delta q(\theta)^\nu \rangle \tau_{3\mu\sigma} \tau_{3\nu\rho} \langle \Delta p(\theta)^\sigma \Delta p(\theta)^\rho \rangle, \tag{2.80}$$

which is

$$\langle (\Delta p)^2 \rangle \langle (\Delta q)^2 \rangle + \langle (\Delta \tilde{p})^2 \rangle \langle (\Delta \tilde{q})^2 \rangle - 2\langle \Delta p \Delta \tilde{p} \rangle \langle \Delta q \Delta \tilde{q} \rangle$$
$$= \langle (\Delta p(\theta))^2 \rangle \langle (\Delta q(\theta))^2 \rangle + \langle (\Delta \tilde{p}(\theta))^2 \rangle \langle (\Delta \tilde{q}(\theta))^2 \rangle$$
$$-2\langle \Delta p(\theta) \Delta \tilde{p}(\theta) \rangle \langle \Delta q(\theta) \Delta \tilde{q}(\theta) \rangle. \tag{2.81}$$

Now let us focus our attention only on the states $|i\rangle$ with the following properties:

(a):   each $|i\rangle$-state is a product of powers of $\alpha^\dagger(\theta)$
       and powers of $\tilde{\alpha}^\dagger(\theta)$ acting on $|0(\theta)\rangle$,

(b):   They are invariant under the tilde conjugation:

$$|i\rangle = |i\rangle^\sim, \qquad \langle i| = \langle i|^\sim. \tag{2.82}$$

Although we derive an uncertainty relation for these states $|i\rangle$, our particular interests lie in the uncertainty relation for the $\alpha(\theta)$-vacuum. Therefore, it is important to note that the $\alpha(\theta)$-vacuum, $|0(\theta)\rangle$, is an example of one of these $|i\rangle$ states.

Since tilde conjugation of a c-number is its complex conjugate and the expectation values of Hermitian operators are real, we have $\langle q \rangle = \langle \tilde{q} \rangle$, $\langle q^2 \rangle = \langle \tilde{q}^2 \rangle$, etc. It is then due to the property (a) of the states $|i\rangle$ that

$$\langle \Delta q(\theta) \Delta \tilde{q}(\theta) \rangle = \langle \Delta q(\theta) \rangle \langle \Delta \tilde{q}(\theta) \rangle = 0, \qquad (2.83)$$

$$\langle \Delta p(\theta) \Delta \tilde{p}(\theta) \rangle = \langle \Delta p(\theta) \rangle \langle \Delta \tilde{p}(\theta) \rangle = 0. \qquad (2.84)$$

The well known uncertainty relation (2.12) is applied to $\Delta p(\theta)$ and $\Delta q(\theta)$:

$$\langle (\Delta q(\theta))^2 \rangle \langle (\Delta p(\theta))^2 \rangle \geq \frac{\hbar^2}{4}, \qquad (2.85)$$

where, in order to avoid confusion, we explicitly write $\hbar$, although it is unity according to our present system of units. Now (2.81) becomes

$$\langle (\Delta q)^2 \rangle \langle (\Delta p)^2 \rangle \geq \frac{\hbar^2}{4} + \langle \Delta q \Delta \tilde{q} \rangle \langle \Delta p \Delta \tilde{p} \rangle. \qquad (2.86)$$

This is the uncertainty relation for the states $|i\rangle$. The first term on the r.h.s. describes the usual quantum fluctuation. Thus this relation shows that the fluctuation in the states $|i\rangle$ is enhanced by the second term, which is

$$\langle \Delta q \Delta \tilde{q} \rangle \langle \Delta p \Delta \tilde{p} \rangle = \frac{\hbar^2}{4} \sinh^2 2\theta \qquad (2.87)$$

when $|i\rangle$ is the α-vacuum $|0(\theta)\rangle$. This implies that the two mode squeezed states contain noise, although they are pure states.

A significant implication of the consideration in this section is the fact that *any bosonic Bogoliubov transformation creates a noise in a pure state*; the same bosonic Bogoliubov transformation creates the same amount of noise *whatever the cause for the transformation* . In the next section we show that this noise is a thermal noise. Therefore, this Bogoliubov transformation is called the thermal Bogoliubov transformation.

It is remarkable that the quantities $\langle \Delta q \Delta \tilde{q} \rangle$ and $\langle \Delta p \Delta \tilde{p} \rangle$ are proportional to $\hbar$ so that the thermal fluctuation would also seem to have a quantum origin. When we derive a variety of macroscopic objects in systems of quantum fields in later sections we will see how this thermal noise can become macroscopic.

2.3.2 CORRELATION THROUGH THE VACUUM

To make our discussion concrete let us work with a simple model in quantum optics. Consider two independent incident modes of light denoted by

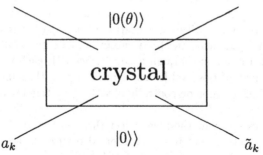

Figure 1. Two beams change their vacuum through a crystal medium, causing a thermal correlation through vacuum.

the two sets of oscillator operators introduced in the last section; $(a, a^\dagger)$ and $(\tilde{a}, \tilde{a}^\dagger)$ (see the Figure 1). Suppose that these modes interact through a crystal medium for a short time. Before the interaction the vacuum of the system is the one annihilated by the operators a and $\tilde{a}$, i.e. $|0\rangle\rangle$. When one considers a type of crystal interaction inducing the Bogoliubov transformation in (2.48) and (2.49), the vacuum changes into $|0(\theta)\rangle$. In other words, the state just at the time when the two waves come out of the crystal is $|0(\theta)\rangle$. This time is called t^1. The time when the waves enter the crystal is denoted by t^0. We are considering the system described by the total Hamiltonian

$$\hat{H} = \hat{H}_0 + \hat{H}_{int}(t), \qquad (2.88)$$

where the $\hat{H}_0$ is the free Hamiltonian

$$\hat{H}_0 = \omega a^\dagger a + \omega' \tilde{a}^\dagger \tilde{a}, \qquad (2.89)$$

and the $\hat{H}_{int}(t)$ is the interaction between a and $\tilde{a}$ mediated by the crystal. We assume that this interaction is the kind which induces the Bogoliubov transformation and therefore, it has the form $g(t)G_B$ with $g(t)$ being a time - dependent coupling constant such that $g(t) = 0$ for $t > t_1$ and $t < t_0$ [14]. The Bogoliubov parameter θ is determined by the coupling constant $g(t)$. The Bogoliubov transformation maps the set $(a, a^\dagger, \tilde{a}, \tilde{a}^\dagger)$ to the set $(\alpha(\theta), \alpha^\dagger(\theta), \tilde{\alpha}(\theta), \tilde{\alpha}^\dagger(\theta))$ according to (2.48) and (2.49). Since $|0(\theta)\rangle$ contains the pair $a^\dagger \tilde{a}^\dagger$ according to (2.64), this state is dependent on time unless $\omega + \omega' = 0$, because the pair term changes in time as $\exp[i(\omega + \omega')(t - t_1)]a^\dagger \tilde{a}^\dagger$; the final vacuum is a state with oscillating pairs.

A remarkable observation in this mechanism is that, as a result of the fact that the two systems share the common vacuum $|0(\theta)\rangle$, a measurement of a manifests noise due to the effect of $\tilde{a}$. This is despite the fact that after interaction the a- and $\tilde{a}$-systems are spatially separated from each other, and therefore have no mutual interactions (i.e. $g(t) = 0$).

To see where this noise comes from it is instructive to recall that the vacuum $|0(\theta)\rangle$ is a state with condensation of bare particles forming $a\tilde{a}$-pairs. Thus, every a-quantum in this vacuum is accompanied by a $\tilde{a}$-quantum. This makes the a- and $\tilde{a}$-quanta *correlated* with each other. As a result, a measurement of observables associated only with a inevitably contains an effect which cannot be controlled by the a-system only, manifesting the noise.

In the above we saw the noise effect through the uncertainty relation in (2.86). This noise effect has been seen also through an expansion of the Wigner function of the bare particles [15, 12].

2.4   Pure State and Mixed State

2.4.1   BOSONIC SYSTEM

In section 2.3 we saw that a noise is created by the Bogoliubov transformation in a pure state. This should surprise us, because we usually expect the appearance of noise only in a mixed state. In this section we show that the same noise created by the Bogoliubov transformation can also be expressed by a statistical average in a mixed state. This will complete the statement that the pure state noise created by the Bogoliubov transformation is indeed thermal noise. To show this we again consider the two oscillators a and $\tilde{a}$. We denote the eigenvalues of $a^\dagger a$ and $\tilde{a}^\dagger \tilde{a}$ by m and $\tilde{m}$ respectively. (Note that $\tilde{m}$ means the eigenvalue m of $\tilde{a}^\dagger \tilde{a}$; the tilde symbol in $\tilde{m}$ indicates that this eigenvalue belongs to the $\tilde{a}$-system and the number beneath the tilde symbol is the eigenvalue. For example m and $\tilde{m}$ both mean the eigenvalue m; one belongs to the nontilde oscillator while another to the tilde oscillator.) We then choose the eigenvectors $|m, \tilde{m}'\rangle$ for basis of the Hilbert space for the two oscillators. Now consider the pure states in this doubled space defined as

$$|0\rangle \;=\; c_1 \sum_{m=0}^{\infty} \rho^\alpha |m, \tilde{m}\rangle, \tag{2.90}$$

$$\langle 0| \;=\; c_2 \sum_{m=0}^{\infty} \langle m, \tilde{m}| \rho^{1-\alpha}, \tag{2.91}$$

where α is any real constant, and c_1 and c_2 are the normalization constants restrained by

$$\langle 0|0\rangle = 1. \tag{2.92}$$

The operator ρ is the density operator defined by

$$\rho = f^{a^\dagger a}, \tag{2.93}$$

where f is a c-number. The normalization condition (2.92) gives

$$c_1 c_2 = 1 - f. \tag{2.94}$$

Exercise 9 Derive (2.94).

The equations (2.90) and (2.91) define the vacuum with a condensation of thermal pairs. We then find for any operator A consisting of a and $a^\dagger$ the relation

$$\langle 0|A(a, a^\dagger)|0\rangle = c_1 c_2 \sum_{m=0}^{\infty} f^m \langle m|A(a, a^\dagger)|m\rangle \tag{2.95}$$

$$= Tr[\rho A(a, a^\dagger)]/Tr\rho. \tag{2.96}$$

Exercise 10 Derive (2.96).
Hint: Use the relation

$$\langle m, \tilde{l}|A(a, a^\dagger)|m', \tilde{l}'\rangle = \langle m|A(a, a^\dagger)|m'\rangle \delta_{ll'}. \tag{2.97}$$

The right hand side of (2.96) is the statistical average of A with respect to the density operator ρ. The relation

$$\begin{aligned} n &= \langle 0|a^\dagger a|0\rangle \\ &= c_1 c_2 \sum_{m=0}^{\infty} m f^m \\ &= \frac{f}{1-f}. \end{aligned} \tag{2.98}$$

relates f to the particle number parameter n. We thus find the equality of the vacuum expectation value (a pure state expectation value in a doubled space) to the statistical average (mixed state expectation value) [16].

According to (2.90) the vacuum $|0\rangle$ is a superposition of the pair states, $|m, \tilde{m}'\rangle$ with $m = m'$. Furthermore, when we choose $\alpha = 1/2$ and $c_1 = c_2^*$, $\langle 0|$ is the Hermitian conjugate of $|0\rangle$. Thus, with these choices of α, c_1 and c_2, the vacuum in (2.90) is related to the $(a, \tilde{a})$-vacuum, that is $|0, \tilde{0}\rangle$, through the thermal Bogoliubov transformation with n the same as the one in (2.98). This shows that the pure state noise considered in this section is created by the thermal Bogoliubov transformation. The problem of two optical beams going through a crystal in section 2.3 showed how noise in a pure state can be experimentally realizable. The equation (2.98) shows that *the vacuum expectation value with $f = \exp[-\beta\omega]$ is identical to the thermal average at the temperature $T = 1/\beta$.* This leads us to a pure-state view for thermal average. Here use was made of the unit for temperature such that $k_B = 1$ for the Boltzmann constant k_B.

2.4.2 THERMAL VACUUM FOR FERMIONIC OSCILLATORS

The consideration in the last subsection (without any change except the values of $c_1 c_2$ and n) can be applied also to two fermionic oscillators:

$$[a, a^\dagger]_+ = 1, \tag{2.99}$$

$$[\tilde{a}, \tilde{a}^\dagger]_+ = 1, \tag{2.100}$$

$$[a, \tilde{a}]_+ = [a, \tilde{a}^\dagger]_+ = 0. \tag{2.101}$$

The equivalence between the vacuum expectation value and thermal average holds true also for these fermionic oscillators. The ket-vacuum and bra-vacuum are defined by (2.90) and (2.91) respectively.

Exercise 11 Show that

$$c_1 c_2 = \frac{1}{1+f} \tag{2.102}$$

and

$$n = \frac{f}{1+f} \tag{2.103}$$

for fermionic oscillators.

Thus the thermal vacuum for fermionic oscillators also contains thermal pairs (that is, $a\tilde{a}$-pairs). Therefore, it should be related to the bare vacuum (that is $|0, \tilde{0}\rangle$) through the fermionic Bogoliubov transformation which mixes a with $\tilde{a}^\dagger$.

In this section we restrict our consideration by requiring that $|0\rangle$ is the Hermitian conjugate of $\langle 0|$ by choosing $\alpha = 1/2$ and $c_1 = c_2^*$. Then the Bogoliubov transformation depends only on the number parameter n. Thus, we denote $|0\rangle$ and $\langle 0|$ by $|0(\theta)\rangle$ and $\langle 0(\theta)|$ respectively in order to indicate explicitly their dependence on parameter θ which determines n.

Since thermal vacua depend on θ as $\langle 0(\theta)|$ and $|0(\theta)\rangle$, we denote their vacuum operators by $\alpha(\theta)$ and $\tilde{\alpha}(\theta)$. The vacuum of $(a, \tilde{a})$ is denoted by $|0\rangle\rangle$. Thus we have

$$a|0\rangle\rangle = \tilde{a}|0\rangle\rangle = 0, \tag{2.104}$$

$$\alpha(\theta)|0(\theta)\rangle = \tilde{\alpha}(\theta)|0(\theta)\rangle = 0. \tag{2.105}$$

Obviously the vacuum operators are required to satisfy the fermionic anti-commutation relations:

$$[\alpha(\theta), \alpha^\dagger(\theta)]_+ = 1, \tag{2.106}$$

$$[\tilde{\alpha}(\theta), \tilde{\alpha}^\dagger(\theta)]_+ = 1, \tag{2.107}$$

$$[\alpha(\theta), \tilde{\alpha}(\theta)]_+ = [\alpha(\theta), \tilde{\alpha}^\dagger(\theta)]_+ = 0. \tag{2.108}$$

Since the transformation is the one mixing a and $\tilde{a}^\dagger$ it has the form

$$\alpha(\theta) = ca + d\tilde{a}^\dagger, \tag{2.109}$$

$$\tilde{\alpha}(\theta) = c\tilde{a} - da^\dagger, \tag{2.110}$$

Here the invariance of the fermionic anticommutation relation under the Bogoliubov translation was considered. This invariance further restricts the c-number parameters c and d as $c^2 + d^2 = 1$. Thus, we can write them as $c = \cos\theta$ and $d = \sin\theta$, respectively.

This transformation is induced by $U_B(\theta)$ in the manner exactly same as the one for bosons:

$$\alpha(\theta) = U_B(\theta)a U_B^{-1}(\theta), \tag{2.111}$$
$$\tilde{\alpha}(\theta) = U_B(\theta)\tilde{a} U_B^{-1}(\theta), \tag{2.112}$$

where $U_B(\theta) = \exp[i\theta G_B]$. The generator G_B is again of the same form as the one for bosons:

$$G_B = i[a\tilde{a} - \tilde{a}^\dagger a^\dagger]. \tag{2.113}$$

The relations (2.109) and (2.110) can be put in the form

$$\left[\begin{array}{c} a \\ \tilde{a}^\dagger \end{array} \right]^\mu = B^{-1}(\theta)^{\mu\nu} \left[\begin{array}{c} \alpha(\theta) \\ \tilde{a}^\dagger(\theta) \end{array} \right]^\nu, \tag{2.114}$$

$$[a^\dagger, \tilde{a}]^\mu = [\alpha^\dagger(\theta), \tilde{\alpha}(\theta)]^\nu B(\theta)^{\nu\mu} \tag{2.115}$$

with the Bogoliubov matrix $B(\theta)$ of the form

$$B(\theta) = \left[\begin{array}{cc} c & d \\ -d & c \end{array} \right]. \tag{2.116}$$

The number parameters are $n \equiv \langle 0(\theta)|a^\dagger a|0(\theta)\rangle$ and $\tilde{n} \equiv \langle 0(\theta)|\tilde{a}^\dagger \tilde{a}|0(\theta)\rangle$. We then have

$$n = \tilde{n} = d^2. \tag{2.117}$$

This gives the parameter $d = \sin\theta$ in terms of the number parameter n.

Exercise 12 Derive (2.117).
   Hint: Make use of the relations (2.109) and (2.110).

Then, we have $|0(\theta)\rangle = U_B(\theta)|0\rangle$, which with the fermionic oscillator variables gives a different result from the bosonic case, namely

$$|0(\theta)\rangle = [\cos\theta + \sin\theta a^\dagger \tilde{a}^\dagger]|0\rangle. \tag{2.118}$$

The following form is useful when we extend this relation to quantum field theory:

$$|0(\theta)\rangle = e^{\ln\cos\theta} e^{\ln[1+a^\dagger\tilde{a}^\dagger \tan\theta]}|0\rangle. \tag{2.119}$$

Exercise 13 Derive (2.118).
   Hint: make use of the relations:

$$G_B|0\rangle = -i\tilde{a}^\dagger a^\dagger|0\rangle, \tag{2.120}$$
$$G_B^2|0\rangle = |0\rangle. \tag{2.121}$$

The relation (2.118) indicates that in $|0(\theta)\rangle$ the $(a\tilde{a})$-pair is condensed. The state $|0(\theta)\rangle$ is called the thermal vacuum for fermions. We can build the $(a, \tilde{a})$-Fock space (denoted by $\mathcal{H}(a, \tilde{a})$) by operations of powers of $a^\dagger$ and $\tilde{a}^\dagger$ on $|0\rangle\rangle$, while the Fock space $\mathcal{H}(\theta)$ is given by operations of $\alpha^\dagger(\theta)$ and $\tilde{\alpha}^\dagger(\theta)$ on $|0(\theta)\rangle$. We see that $\mathcal{H}(a, \tilde{a})$ is equivalent to $\mathcal{H}(\theta)$ and therefore that the parameterized Fock spaces $\mathcal{H}(\theta)$ forms an equivalent set. Since the $\alpha(\theta)$-vacua are the states with condensation of $a\tilde{a}$-pairs, they are invariant under exchange of a and $\tilde{a}$. This exchange is the tilde conjugation.

2.4.3   NON-TRIVIAL PARAMETERS IN VACUA

The equivalence between the vacuum expectation value and ensemble average shown in (2.96) is based on the definitions of the vacua in (2.90) and (2.91). Therefore, the parameters which appear in this definition of vacua are the parameters which show up in pure state formulation for thermal noise.

The obvious trivial parameter is the phase factors which can be absorbed by a or $\tilde{a}$. This is trivial because this freedom appears in any quantum mechanical system when we make use of the oscillator representation. The phase factor of c_1, which determines the phase factor of c_2 as $c_1 c_2$, is also trivial.

There is a nice way of getting rid of these trivial freedoms. In section 2.2 we called the exchange of tilde and non tilde operator the tilde conjugation. There it was shown that the vacuum is invariant under the tilde conjugation. In subsection 2.3.1 we found it convenient to extend the definition of the tilde conjugation in such a manner that it includes the replacement of any c-number with its complex conjugate. With this definition of tilde conjugation, we preserve the requirement that the vacuum is invariant under the tilde conjugation. This eliminates the trivial phase freedoms mentioned above. Now c_1 and c_2 are real.

However there are three parameters which are not trivial. One is α-freedom which originates from the identity $Tr[\rho A] = Tr[\rho^{1-\alpha} A \rho^\alpha]$. Another freedom which is not trivial is the ratio (c_1/c_2). These two parameters are not trivial because they can ensure that the bra-vacuum is not the Hermitian conjugate of the ket-vacuum as (2.90) and (2.91) show.

If we require that the bra-vacuum is the Hermitian conjugate of the ket-vacuum, these parameters are then fixed to be

$$\alpha = \frac{1}{2} \tag{2.122}$$

and

$$\frac{c_1}{c_2} = 1. \tag{2.123}$$

The pure state formalism with this choice is called the unitary representation. When we discuss noise in pure states in this and the following chapter, we use the unitary representation. However, in thermal quantum field theories, the non unitary choices are frequently used. This is not because α and (c_1/c_2) have any physical relevance, but because using the freedom helps to avoid technical computational problems.

The parameter with real physical significance is f which is determined by the density matrix ρ according to (2.93). This is related to the number parameter n as

$$n = \frac{f}{1 - \sigma f}. \tag{2.124}$$

where $\sigma = +1$ (-1) for bosons (fermions).

2.5   TFD Mechanism for Noise Creation

The appearance of thermal noise in pure states makes the idea of formulating a thermal physics in terms of pure states seem promising. This suggests an even more challenging task of formulating a thermal theory for isolated systems (an isolated system is one which is completely isolated, e.g. the whole universe). However, the discussions of the last sections cannot be applied to thermal physics, as long as a and $\tilde{a}$ are dynamical operators; since both ω and ω' are positive for dynamical systems, those considerations cannot be used in the description of a stationary equilibrium system. Indeed, in pure states of systems with dynamical degrees of freedom, there is no degree of freedom left to account for the thermal degree of freedom. The thermal degree of freedom is something we find only when the thermal situation changes. If the temperature were the same all the time everywhere, then we would never discover the thermal degrees of freedom. This freedom of changing the thermal situation is not a part of the dynamical freedom. Thus, in a pure state formalism of thermal physics, the thermal degree of freedom should be a new freedom added to the dynamical ones. The TFD mechanism for creation of noise in a pure state discussed in the last section suggests that the thermal degree of freedom might be obtained by doubling each dynamical degree of freedom in such a manner that, to a dynamical variable a is associated its tilde conjugate $\tilde{a}$; this doubling degrees of freedom provides us with the freedom of making the Bogoliubov transformation. In thermal physics there are many observable thermal quantities such as heat energy, entropy, etc. and there exist many thermal processes such as phase transitions. Therefore, it is an interesting question to ask if we can describe these thermal quantities and thermal transitions in terms of pure state concepts by doubling the degrees of freedom

However doubling the degrees of freedom is not sufficient for thermal physics. To be able to deal with many different macroscopic ordered states, a theory should be able to accommodate a large variety of macroscopic ordered states. This is not possible for quantum mechanics with finite degrees of freedom, suggesting the need for a quantum field theory.

There is another simple reason for using field theory. In conventional thermal physics, thermal control is done through a heat bath which should carry infinite capacity or infinite degrees of freedom, that is a field. Therefore, a pure state formulation may need doubling of dynamical degrees of freedom in the framework of quantum field theory. Such a formalism is called thermo field dynamics (TFD). This will be explained in later chapters.

However, before we enter thermal problems, we should understand how the quantum field theory deals with macroscopic phenomena. This will be considered in the next chapter.

2.6   TFD Mechanism and Black Hole

There is a dynamical system which creates thermal effects in our world through the TFD mechanism. That is the black hole. Since this is obviously a subject of gravitational field theory, it does not really belong to the subject of this chapter which is devoted to quantum mechanics. However a similarity between the thermal noise created by a black hole and the noise in two mode squeezed state is so remarkable that it is very briefly discussed here.

In two subsections, 2.3.1 and 2.3.2, it was shown how a correlation through the vacuum creates thermal noise in two systems without any mutual interaction. This was called the TFD mechanism. The same situation can be seen in the physics of a black hole. A black hole creates a horizon. The world beyond the horizon is completely dissociated from our world. However, these two dissociated worlds share a common vacuum. Let a stand for dynamical variables of our world, while those in the world beyond the horizon are represented by $\tilde{a}$. It has been shown that the vacuum is not the vacuum, $|0\rangle\rangle$ associated with a and $\tilde{a}$, but is $|0(\theta)\rangle\rangle$, which is related to $|0\rangle\rangle$ through the thermal Bogoliubov transformation. The parameter θ of the Bogoliubov transformation is determined by the mass of the black hole, while it determines the temperature of radiation around the black hole (the Hawking temperature) [17, 18]. Thus the analogy with the case of a two mode squeezed state is perfect. Furthermore, it has been shown that the Hamiltonian which controls the entire system has the structure $\hat{H} = H - \tilde{H}$, in which H is the Hamiltonian for dynamical phenomena in our world and $\tilde{H}$ is obtained from H by the tilde conjugation rules. For the free oscillator model in section 2.3.2 this means $\omega + \omega' = 0$, implying the stationary nature of the vacuum $|0(\theta)\rangle\rangle$.

An extension of this picture motivates an idea that the origin of the thermal degree of freedom in our world could be due to the presence of other universes which are totally dissociated from our world, though they share the vacuum with our world [19]. This subject will not be pursued in this book any further. We only note that this description of the thermal situation around a black hole is exactly the same as the TFD formalism for equilibrium states.

2.7 Phase and Time Operators in TFD

In the previous sections we showed how the doubled oscillators, a and $\tilde{a}$, treat a mixed state average associated with the a-oscillator in terms of pure state of doubled oscillators. In this section we discuss the phase operator and time operator in this doubled oscillator formalism. In subsection 2.2.1 we introduced a problem of constructing the phase operator. The main cause of the difficulty is that the number spectrum is confined to the non-negative domain, and therefore, it is lower bounded. It was shown by Newton in [4] that, when we introduce the lower unbounded number by doubling degrees of freedom, a phase operator appears. This consideration has recently been elaborated by Ban in the article [6], and applied to the TFD mechanism. The time operator was also discussed in this article. In this section we study this approach, closely relying on this article.

We saw that the Hamiltonian for the stationary vacuum of the two mode squeezed system has the form

$$\hat{H}_0 \;=\; H_0 - \tilde{H}_0 \tag{2.125}$$
$$\;=\; \omega \hat{N} \tag{2.126}$$

with the number operator

$$\hat{N} = a^\dagger a - \tilde{a}^\dagger \tilde{a}. \tag{2.127}$$

We have seen that the mixed state average associated with the a-system is given by the pure state vacuum expectation value associated with the $(a, \tilde{a})$-system. Thus, the energy operator of the a-system is H_0 whose eigenvalue is non-negative, but the eigenvalue of the Hamiltonian $\hat{H}_0$ is lower unbounded. How this formalism is applied to thermal field theories will be explained in the later chapters. Here, it is pointed out that the canonical conjugate of $\hat{H}_0$ does exist and acts as the time operator. In the same way the canonical conjugate of the number operator $\hat{N}$ exists and it acts as a phase operator.

Let us begin with the phase operator. We can choose the base set of the Hilbert space by the complete set of the vectors $|n, \tilde{m}\rangle\rangle$, in which n and $\tilde{m}$ stand for the eigenvalues of N and $\tilde{N}$, respectively. Each member of this set can be specified by the eigenvalues of $\hat{N}$ denoted by $\hat{n}$ together with $\tilde{m}$. We denote this state by $|\hat{n}, \tilde{m}\rangle\rangle$. The base set of the Hilbert space is $(|\hat{n}, \tilde{m}\rangle\rangle :$ $-\infty < \hat{n} < \infty, 0 \leq \tilde{m} < \infty)$. We then have the orthonormalization relations

$$\langle\langle \hat{n}, \tilde{m} | \hat{n}', \tilde{m}' \rangle\rangle = \delta_{nn'} \delta_{mm'} \tag{2.128}$$

and

$$\sum_{m=0}^{\infty} \sum_{n=-\infty}^{\infty} |\hat{n}, \tilde{m}\rangle\rangle\langle\langle \hat{n}, \tilde{m} | = 1, \tag{2.129}$$

where 1 means the identity operator.

Introduce the operator D by the relation

$$D = \sum_{m=0}^{\infty} \sum_{n=-\infty}^{\infty} |\hat{n} - 1, \tilde{m}\rangle\rangle\langle\langle\hat{n}, \tilde{m}|. \tag{2.130}$$

We can then prove the relations

$$DD^{\dagger} = D^{\dagger}D = 1, \tag{2.131}$$
$$D|\hat{n}, \tilde{m}\rangle\rangle = |\hat{n} - 1, \tilde{m}\rangle\rangle, \tag{2.132}$$
$$D^{\dagger}|\hat{n}, \tilde{m}\rangle\rangle = |\hat{n} + 1, \tilde{m}\rangle\rangle. \tag{2.133}$$

Exercise 14 Derive the above relations.

The above relations lead us to

$$[D, \hat{N}] = D, \tag{2.134}$$

which shows that D is the exponential of ($i \times$ phase operator). To clarify this point, we introduce the states

$$|\theta, \tilde{m}\rangle\rangle = \frac{1}{\sqrt{2\pi}} \sum_{n=-\infty}^{\infty} \exp\left(-in\theta\right)|\hat{n}, \tilde{m}\rangle\rangle. \tag{2.135}$$

Then, making use of the relation

$$\sum_{n=-\infty}^{\infty} \exp\left(-in\theta\right) = 2\pi\delta(\theta), \tag{2.136}$$

with the θ being confined to the domain $-\pi \leq \theta \leq \pi$, we can derive the orthonormalization relations, i.e.

$$\langle\langle\theta, \tilde{m}|\theta', \tilde{m}'\rangle\rangle = \delta_{mm'}\delta(\theta - \theta') \tag{2.137}$$

and

$$\sum_{m=0}^{\infty} \int_{-\pi}^{\pi} d\theta\, |\theta, \tilde{m}\rangle\rangle\langle\langle\theta, \tilde{m}| = 1. \tag{2.138}$$

Exercise 15 Derive the relation (2.136).

Hint1: Prove that the left hand side vanishes with nonvanishing θ.

Hint2: Integrate the left hand side over the entire domain of θ, and show that the terms with nonvanishing n have no contribution.

Now we can derive the relation

$$D|\theta, \tilde{m}\rangle\rangle = \exp\left(-i\theta\right)|\theta, \tilde{m}\rangle\rangle. \tag{2.139}$$

Thus, we can write

$$D = \sum_{m=0}^{\infty} \int_{-\pi}^{\pi} d\theta \, |\theta, \tilde{m}\rangle\rangle \exp\left(-i\theta\right)\langle\langle\theta, \tilde{m}|. \qquad (2.140)$$

We thus obtained the *phase operator* conjugate to $\hat{N}$. By replacing $\hat{N}$ with $\hat{H}_0$, we obtain the *time operator*.

In chapters 7, 8 and 9 we develop a real time formalism of thermal quantum field theories which is called the thermo field dynamics (TFD). There we will see that the consideration in this section can be applied straightforwardly to TFD. Thus, we find that there are phase and time operators in TFD. We have not fully appreciated the implication of this result in understanding and evaluating TFD.

2.8 REFERENCES

[1] L. Susskind and J. Glogower, *Physics* **1** (1964) 49.

[2] P. Carruthers and M. M. Nieto, *Rev. Mod. Phys.* **40** (1968) 411.

[3] J. M. Levy-Leblond, *Ann. Phys.* **101** (1976) 319.

[4] R. Newton, *Ann. Phys.* **124** (1980) 327.

[5] D. T. Pegg and S. M. Barnett, *Phys. Rev.* **A41** (1990) 3427.

[6] M. Ban, *Journal Math. Phys.* **32** (1991) 3077.

[7] R. J. Glauber, *Phys. Rev.* **131** (1963) 2766.

[8] B. Yurke and M. Potasek, *Phys. Rev.* **A36** (1987) 3464.

[9] S.M. Barnett and P.L. Knight, *J. Opt. Am.* **B2** (1985) 467.

[10] S.M. Barnett and P.L. Knight, *Phys. Rev.* **A38** (1988) 1657.

[11] A. Mann and M. Revzen, *Phys. Lett.* **134A** (1989) 273.

[12] Y.S.Kim and Ming Li, *Phys. Lett.* **139A** (1989) 443.

[13] A. Mann, M. Revzen, H. Umezawa, and Y. Yamanaka, *Phys. Lett.* **140A** (1989) 475.

[14] A. Mann and M. Revzen, *Phys. Lett.* **134A** (1988) 531.

[15] M. Berman, *Phys. Rev.* **A40** (1989) 2057.

[16] Y. Takahashi and H. Umezawa, *Collect. Phenom.* **2** (1975) 55.

[17] W. Israel, *Phys. Lett.* **57A** (1976) 107.

[18] H. Umezawa, A.E.I. Johansson, and Y. Yamanaka, *Class. Quantum Grav.* **7** (1990) 385.

[19] R. Laflamme, *Physica* **158A** (1989) 58.

3
Inequivalent Vacua

3.1 Introduction

Quantum mechanics opened a door to the microscopic world. However, the discovery of electron energy level shifts in atoms (the Lamb shift) due to radiative corrections showed that quantum mechanics does not cover the entire microscopic world. The need of radiative correction indicates that quantum field theory is required. According to present day physics the fundamental world consists of quark fields, lepton fields and many kinds of gauge fields. Therefore, it is natural that more and more phenomena will require quantum field theory for their description. This situation will continue until new physics beyond quantum field theory emerges.

In the early stages of the development of quantum field theory, it was widely felt that quantum field theory was a straightforward extension of quantum mechanics to systems with a very large number of degrees of freedom. Physicists tended to accept that most of the laws known in quantum mechanics applied in quantum field theory too. This includes the law of equivalence of all representations which was explained in the last chapter. However such a simple minded view was upset by new phenomena which suggested the need for a drastically new viewpoint. In this section some of them will be discussed briefly.

3.1.1 INFRARED CATASTROPHE OF BREMSSTRAHLUNG

The development of quantum field theory started with quantum electrodynamics in 1929. Therefore, most of the early applications were calculations of electromagnetic quantities. The calculation method was a straightforward extension of the perturbation scheme formulated for quantum mechanics. Thus the Hamiltonian was put in the form $H = H_0 + H_{int}$ with unperturbed Hamiltonian H_0 and the calculation was formulated in expansion in powers of the interaction Hamiltonian H_{int}. Among many problems Compton scattering and Bremsstrahlung by electrons attracted the attention of many physicists. However, the lowest order perturbation calculation of Bremsstrahlung immediately met with a catastrophe; the transition cross section diverged due to low energy photon contributions. The difficulty was resolved by F. Bloch and A. Nordsieck [1]. They showed by means of a canonical transformation that the observed electron carries a cloud of photons and that the infrared catastrophe disappears when we consider the soft photon part of this cloud. We can understand this situation intuitively as follows [2]. Let us assume a random distribution of photons around an electron. The random distribution assumption is reasonable when we consider only the soft photons, because soft photons cause practically no change in momentum when interacting with electrons. Let a denote the probability of finding one photon. Then, the probability of finding n photons is $(1/n!)a^n$. This leads to the Poisson distribution of photons and introduces the probability normalization factor $\exp(-a)$ in order to make the total probability unity. Thus the probability of finding n photons becomes $(1/n!)a^n \exp(-a)$. The infrared catastrophe was caused by the fact that a is infinite. This now makes the probability for finding any finite number of soft photons vanish, so that *an infinite number of soft photons are present around each electron.* An electron surrounded by an infinite number of photons can never be obtained by any finite order perturbation calculation as only a finite number of particles are taken into account when the unperturbed state is the bare electron, i.e. the electron without photon.

The fact that the probability for finding a finite number of photons is zero has a dramatic implication; it shows that the state vector of the observed electron cannot be expanded by state vectors in the Hilbert space for bare electrons and photons in a meaningful manner. In other words, *the observed electron state vector does not seem to belong to the Hilbert space for bare electrons and photons.* In the last chapter we saw that in quantum mechanics all of the representations are equivalent to each other. The above analysis of structure of observed electrons suggests that this equivalence theorem may not hold in quantum field theory. However, the significance of this aspect of the problem of the infrared catastrophe was missed by the majority of physicists at that time.

The study of infrared catastrophe in calculations of Bremsstrahlung cross sections showed that observed particles are very different from the bare particles.

3.1.2 VACUUM POLARIZATION

It is well known in classical electromagnetism that the effect of an external field is modified in an electromagnetic medium. When an electric charge is put in a medium it attracts charges of the opposite sign and repels those with the same sign. This screening weakens the charge. It is therefore interesting to see how external electromagnetic fields and charges behave in the vacuum of the quantum field theory which is full of virtual particles and acts like a medium. When we introduce an external effect with the vector potential $\vec{A}$, it interacts with the current $\vec{j}$ of the quantum field through the interaction Hamiltonian $\vec{j} \cdot \vec{A}$. The current it induces is

$$J_i = -i \int_{-\infty}^{t} dt' \int d^3x' \langle 0|[j_i(\vec{x},t), j_l(\vec{x}',t')]|0\rangle A_l(\vec{x}',t'), \qquad (3.1)$$

which is called the response formula. With this formula we can calculate the change in the charge, current, electromagnetic field and in the vector potential for the photon field. If the charge should be weakened, the observed charge e_r should be smaller than the original charge e_0 (i.e. bare charge). Writing as $e_r = Z_3^{1/2} e_0$ with use of the common notation, Z_3 would be unity if there is no change, while $1 > Z_3 \geq 0$ if it is weakened. With use of the response formula it was proven without use of perturbation calculation and without specifying a model that $1 > Z_3 \geq 0$ [3]. This phenomenon is called vacuum polarization. The change in the charge is called charge renormalization.

In the early days of renormalization, physicists used to associate charge renormalization only with the charged particle under consideration; for example, calculations of the renormalized electron charge only considered the effect of vacuum electrons. This naturally led to a puzzling result that the renormalized charge of the electron differed from that of the μ-meson. This problem is resolved when we realize that all charged particles contribute to the vacuum polarization. This can be better understood when we find that the charge renormalization is associated with the change of the vector potential of the photon. Denoting the bare vector potential by A and modified (i.e. renormalized) vector potential by A_r, it can be shown that $A_r = Z_3^{-1/2} A$. This is called the wave function renormalization of the photon. Since the photon interacts with all of the charged particles, its renormalization causes the charge renormalization. S. Tomonaga summarized (private conversation) this fact by the simple and clear statement that "the charge belongs, not to the charged particles, but to the photon".

This renormalization effect shows the remarkable fact that the vacuum is a materially rich medium. The vacuum polarization makes the observed photon contain, not only the bare photon, but also an infinite number of electrons, positrons and other charged particles. The wave function renormalization induces a puzzling question, asking if the observed photon field theory would then violate the canonical commutation relations which are fundamental to quantum theory. This question is resolved without violating the canonical commutation relation when we find that the Hilbert space for observed particles is not equivalent to the bare ones, meaning that the interaction effect cannot be treated by a transformation confined to one Hilbert space.

It should be noted that the above feature of observed charge smaller than the bare one is not universal in gauge theories. There are some non-Abelian gauge theories in which the observed coupling constant is larger than the bare one. In this case, when we penetrate inside an observed particle, the coupling constant becomes weaker and we find the theory so called asymptotically free.

3.1.3 RENORMALIZATION

In the above we have seen several phenomena which suggest that the observed particles are very different from the bare particles and that the state vectors of observed particles might not belong to the Hilbert space for bare particles. The mapping from the bare particle parameters to the physical ones is called the renormalization. Since we should stick to only one Hilbert space and this space should be the one for observed states, we should use only the Fock space for the observed particles. In the terminology of perturbation theory this means that the unperturbed representation should be the one with observed parameters (such as the observed mass) of particles. The coupling constants should also be expressed in terms of observed parameters. This is the renormalized theory. We have mass renormalization, coupling constant renormalization and wave function (or field) renormalization. It is well known that we can eliminate ultraviolet divergence in relativistic field theory through the renormalization procedure. However, the above consideration shows that *the renormalization should be performed even in cases such as condensed matter physics where there is no ultraviolet divergences.*

These phenomena cast doubt on the validity of the equivalence theorem among the Hilbert spaces in quantum field theory. The equivalence theorem in quantum mechanics was illustrated in the last chapter by analyzing the coherent or squeezed states. Therefore, in the following we test the equivalence theorem in quantum field theory by using the same examples extended to quantum field theory.

3.2 Many Vacua

We begin this study of the vacuum in quantum field theory by considering the Bogoliubov transformation for coherent states. We reproduce the consideration of section 2.2 by means of a quantum field system. We therefore add the momentum suffix to each oscillator operator and consider either one set a_k or two sets $(a_k, \tilde{a}_k)$ of oscillator operators. Let us begin with the coherent state for a_k. We list the relations which are obtained from those in 2.2;

$$[a_k, a_l^\dagger] = \delta(\vec{k} - \vec{l}), \tag{3.2}$$

$$a_k|0\rangle\rangle = 0, \tag{3.3}$$

$$a_k \to \alpha_k(\theta) = a_k + \theta_k, \tag{3.4}$$

$$\alpha_k(\theta)|0(\theta)\rangle = 0. \tag{3.5}$$

We then have

$$a_k|0(\theta)\rangle = -\theta_k|0(\theta)\rangle, \tag{3.6}$$

$$G_c(\theta) = -i \int d^3k \, (\theta_k^* a_k - \theta_k a_k^\dagger), \tag{3.7}$$

$$U_c(\theta) = \exp\left[iG_c(\theta)\right], \tag{3.8}$$

and

$$
\begin{aligned}
|0(\theta)\rangle &= U_c(\theta)|0\rangle \tag{3.9} \\
&= \exp\left(-\frac{1}{2}\int d^3k\,|\theta_k|^2\right)\exp\left(-\int d^3k\,\theta_k a_k^\dagger\right)|0\rangle. \tag{3.10}
\end{aligned}
$$

Exercise 1 Applying the derivation of the relation (2.24) in chapter 2 to the quantum field system under consideration, derive (3.10).

For simplicity let us confine our attention to those cases in which the vacuum $|0(\theta)\rangle$ is invariant under the spatial translation. Then θ_k is the Fourier amplitude of a constant independent of spatial position, implying that it has the form $\theta\delta(\vec{k})$, which gives

$$\int d^3k\,|\theta_k|^2 = \infty. \tag{3.11}$$

In other words,

$$\exp\left(-\frac{1}{2}\int d^3k\,|\theta_k|^2\right) = 0, \tag{3.12}$$

though

$$\langle 0(\theta)|0(\theta)\rangle = 1, \tag{3.13}$$

which follows from the relation

$$\langle 0(\theta)| = \langle\langle 0|U_c^{-1}(\theta). \tag{3.14}$$

These relations together with (3.10) lead to

$$\langle\langle m_k|0(\theta)\rangle = 0 \tag{3.15}$$

for any finite numbers m_k. Here use was made of the notation

$$|m_k\rangle\rangle = \frac{1}{\sqrt{m_k!}}(a_k^\dagger)^{m_k}|0\rangle\rangle. \tag{3.16}$$

This indicates that the probability of finding any state of finite number of a-particles in $|0(\theta)\rangle$ is vanishing, and therefore, that $|0(\theta)\rangle$ contains infinite a-particles. This is a situation exactly the same as the one for the observed electron which we discussed when we studied the infrared catastrophe in the Bremsstrahlung. *The equation (3.10) shows explicitly that the α-vacuum $|0(\theta)\rangle$ cannot be a superposition of state vectors which belong to the space $\mathcal{H}(a)$ which is built on the basis vectors made by cyclic operations of $a^\dagger$ on $|0\rangle$.* This is true for any vector in the space $\mathcal{H}(\theta)$ which is built on the basic vectors made by cyclic operations of $\alpha(\theta)^\dagger$ on $|0(\theta)\rangle$. We see that $\mathcal{H}(a)$ is *inequivalent* to $\mathcal{H}(\theta)$ in the sense that any vector in $\mathcal{H}(a)$ is not a superposition of vectors in $\mathcal{H}(\theta)$ and vice versa. We have the parameterized Fock space $\mathcal{H}(\theta)$ such that any two spaces with different θ are *inequivalent* to each other. This shows how the equivalence theorem for Hilbert spaces in quantum mechanics breaks down in quantum field theory.

Here the following note is in order because of its vital significance. Although the vacuum $|0\rangle\rangle$ of a_k does not belong to the Fock space $\mathcal{H}(\theta)$ with nonvanishing θ, the operators a_k are well defined in the sense that their action on state vectors in $\mathcal{H}(\theta)$ is defined by the relation in (3.4). Since the operators a_k do not annihilate vacuum $|0(\theta)\rangle$, when we use the Fock space $\mathcal{H}(\theta)$, they cannot be called the annihilation operators, even though they satisfy the oscillator commutation relations (3.2). The moral is that, although the relation (3.10) for the structure of the vacuum $|0(\theta)\rangle$ might help us to cultivate our intuition, a mathematical analysis should not depend on such a relation. It is not mathematically sensible because the coefficient in (3.10) vanishes. A mathematical analysis should be based on mathematically meaningful relations such as the relations (3.4) and (3.5).

The above argument shows that the reason for the inequivalence among the Fock spaces above is the condensation of an infinite number of the a-particles in the vacuum $|0(\theta)\rangle$. When we assume the invariance of the vacuum under spatial translation, then in order for two vacua to behave differently when observed there should be a finite density of particles in condensation, leading to an infinite number of particles in the condensate of a system of infinite volume. Indeed, since $(2\pi)^3[\delta(\vec{k})]_{\vec{k}=0}$ in the above argument is the spatial integration of $\exp[i\vec{k}\vec{x}]$ with $\vec{k}=0$, it is V which is the infinite volume. This might give the wrong idea that infinite volume is the origin for appearance of many Fock spaces. The discussion in the next chapter will show that the origin of the appearance of many inequivalent Fock space is the infinite number of degrees of freedom carried by quantum fields.

A similar situation occurs in case of the Bogoliubov transformation for the squeezed states. We now consider a generator consisting of two commuting sets of oscillators:

$$[a_k, a_l^\dagger] = [\tilde{a}_k, \tilde{a}_l^\dagger] = \delta(\vec{k} - \vec{l}), \tag{3.17}$$

$$[a_k, \tilde{a}_l] = [a_k, \tilde{a}_l^\dagger] = 0. \tag{3.18}$$

We extend the consideration for two mode squeezed states in section 2.2 to these quantum fields. Thus, the generator is

$$G_B(\theta) = i \int d^3k \, \theta_k [a_k \tilde{a}_k - \tilde{a}_k^\dagger a_k^\dagger]. \tag{3.19}$$

The transformation is

$$\alpha_k(\theta) = c_k a_k - d_k \tilde{a}_k^\dagger, \tag{3.20}$$

$$\tilde{\alpha}_k(\theta) = c_k \tilde{a}_k - d_k a_k^\dagger. \tag{3.21}$$

with $c_k = \cosh\theta_k$ and $d_k = \sinh\theta_k$ where θ_k is the transformation parameter.

The a-vacuum and α-vacuum are defined by

$$a_k|0\rangle\rangle = \tilde{a}_k|0\rangle\rangle = 0, \qquad (3.22)$$

$$\alpha_k(\theta)|0(\theta)\rangle = \tilde{\alpha}_k(\theta)|0(\theta)\rangle = 0. \qquad (3.23)$$

Then, we have

$$|0(\theta)\rangle = U_B(\theta)|0\rangle\rangle \qquad (3.24)$$

$$= \exp\left[-\delta^{(3)}(0)\int d^3k \ln\cosh\theta_k\right]$$

$$\times \exp\left[\int d^3k \, [a_k^\dagger \tilde{a}_k^\dagger \tanh\theta_k\right]|0\rangle\rangle \qquad (3.25)$$

with $U_B(\theta) = \exp[i\theta G_B]$. Here $\delta^{(3)}(0)$ is $\delta(\vec{k})$ with $\vec{k} = 0$. This indicates that in $|0(\theta)\rangle$ the $(a_k\tilde{a}_k)$-pairs are condensed. The state $|0(\theta)\rangle$ is called the thermal-like vacuum and $(a_k\tilde{a}_k)$-pairs are called the thermal pairs. We can build the basis vectors of the $(a_k, \tilde{a}_k)$-Fock space (denoted by $\mathcal{H}(a,\tilde{a})$) by cyclic operations of $(a_k^\dagger, \tilde{a}_k^\dagger)$ on $|0\rangle\rangle$, while the Fock space $\mathcal{H}(\theta)$ is built on the basic vectors made by cyclic operations of $(\alpha_k(\theta)^\dagger, \tilde{\alpha}_k(\theta)^\dagger)$ on $|0(\theta)\rangle$. Since $\delta^{(3)}(0) = \infty$, the equation (3.25) shows that $\mathcal{H}(a,\tilde{a})$ is inequivalent to $\mathcal{H}(\theta)$. It is due to the same reason that two spaces $\mathcal{H}(\theta)$ and $\mathcal{H}(\theta')$ are inequivalent to each other when $\theta \neq \theta'$. Therefore, the parameterized Fock spaces $\mathcal{H}(\theta)$ form a set of inequivalent spaces.

As was discussed in the case of the coherent state, the generator for the squeezed state, G_B is not mathematically well founded. The relations (3.20) and (3.21) are still well defined and the operations of a_k and $\tilde{a}_k$ on vectors in $\mathcal{H}(\theta)$ are defined through these relations. However, since a_k and $\tilde{a}_k$ do not annihilate the vacuum $|0(\theta)\rangle$, they are not the annihilation operators.

The number parameter n_k is defined by

$$n_k\delta(\vec{k} - \vec{l}) \equiv \langle 0(\theta)|a_k^\dagger a_l|0(\theta)\rangle. \qquad (3.26)$$

This gives $n_k[\delta(\vec{q})]_{\vec{q}=0} = \langle 0(\theta)|a_k^\dagger a_k|0(\theta)\rangle$ which is equal to $\langle 0(\theta)|\tilde{a}_k^\dagger \tilde{a}_k|0(\theta)\rangle$. Since $(2\pi)^3[\delta(\vec{q})]_{\vec{q}=0} = V$, $(2\pi)^{-3}n_k$ is the number density. In the following the n_k will be simply called the number density or number parameter.

In this section we saw that in quantum field theory there are continuous sets of inequivalent vacua, leading to continuous sets of inequivalent Fock spaces. In section 4.3 we will present a general argument as to why many inequivalent Fock spaces appear in quantum field theory.

3.3   Anomalous Operators

Since we frequently make use of such generators as $G_c(\theta)$ or $G_B(\theta)$ although they are mathematically unfounded, we call them anomalous operators. If one really wishes to insist a use of the operator $G_c(\theta)$, we first replace θ_k as

$$\theta_k \rightarrow \theta_k f(\vec{k}) \qquad (3.27)$$

with the square integrable function $f(\vec{k})$:

$$\int d^3k |f(\vec{k})|^2 < \infty. \qquad (3.28)$$

Then the pathological nature of G_c does not appear as long as $f(\vec{k})$ is square integrable. Finally the limit $f(\vec{k}) \rightarrow \delta(\vec{k})$ is to be performed. We usually skip mentioning of this smearing out trick for anomalous operators. In the same way $G_B(\theta)$ in (3.19) may be replaced by

$$G_B(\theta) = i \int d^3k \int d^3q \, \theta_k f(\vec{q}) [a_{k+q}\tilde{a}_{-k} - \tilde{a}^\dagger_{-k} a^\dagger_{k+q}] \qquad (3.29)$$

with the above square integrable function $f(\vec{q})$.

This smearing out trick is formulated in the spatial representation as follows. Consider a generator of the form

$$\int d^3x \rho(\vec{x}), \qquad (3.30)$$

which is assumed to be anomalous. The smearing out trick replaces this with

$$\int d^3x f(\vec{x})\rho(\vec{x}) \qquad (3.31)$$

with a square integrable function

$$\int d^3x |f(\vec{x})|^2 < \infty. \qquad (3.32)$$

The $f(\vec{k})$ is the Fourier transform of $f(\vec{x})$.

3.4   Pure State and Mixed State

3.4.1   BOSONIC SYSTEMS

In section 2.3 we saw that noise is created in a pure state by the Bogoliubov transformation. In section 2.4 we showed that the same noise created by the Bogoliubov transformation can also be expressed by a statistical average in a mixed state. By introducing the momentum suffix, we may try to extend the whole of these considerations to quantum field theory. We need to proceed carefully in doing this. For example, when we apply the uncertainty relation (2.86) in section 2.3 to the canonical operators

$(\vec{q}_k, \vec{p}_k)$ carrying the momentum suffix k, both the right and left sides are proportional to $[V/(2\pi)^3]^2$, where $V = (2\pi)^3\delta(\vec{k})$ with $\vec{k} = 0$. This is not a problem; it reflects the fact that the plane wave states are not normalizable, and therefore, they do not belong to the Fock space. We should derive the uncertainty relations for the wave packet operators given by

$$p_i = \int d^3k f_i(\vec{k}) p_k \tag{3.33}$$

$$q_i = \int d^3k f_i(\vec{k}) q_k. \tag{3.34}$$

Here, $f_i(\vec{k})$ are the members of an orthonormalized complete set of square integrable functions. In this way the discussion of noise in section 2.3 based on the uncertainty relation (2.86) can be reproduced. A touchy point appears also in the study of the relation between the pure state expectation values and mixed state averages. When we try to extend the relation (2.96) for any operator A consisting of a_k and $a_k^\dagger$ to the quantum field system by writing as

$$\langle 0|A(a, a^\dagger)|0\rangle = Tr[\rho A(a, a^\dagger)]/Tr\rho, \tag{3.35}$$

definition of the trace and derivation of this relation require several expressions in terms of the bare states which are not mathematically sensible when we talk about the space $\mathcal{H}(\theta)$, although this provides us with an intuitive understanding of the relation between pure state noise and noise in statistical mechanics. Therefore, in the following, without going through statistical mechanics we try to extend the pure state formulation of the last chapter directly to quantum field theory. The pure state noise in squeezed states for two sets of quantum field oscillators provides a good preparation for thermo field dynamics which will be developed in later chapters.

The pure state formulation for the Hawking radiation around the black hole discussed in section 2.6 is intrinsically a field theoretical one, so the formulation in terms of the field theory is the sensible one.

3.4.2 FERMIONIC SYSTEMS

It was shown in section 2.4.2 that the discussion in the previous section 3.2 can easily be extended to the thermal vacuum of fermionic systems:

$$[a_k, a_l^\dagger]_+ = \delta(\vec{k} - \vec{l}), \tag{3.36}$$

$$[\tilde{a}_k, \tilde{a}_l^\dagger]_+ = \delta(\vec{k} - \vec{l}), \tag{3.37}$$

$$[a_k, \tilde{a}_l]_+ = [a_k, \tilde{a}_l^\dagger]_+ = 0. \tag{3.38}$$

We only sketch a brief summary.

The thermal vacuum for fermionic oscillators also contains the thermal pairs (that is, $a_k\tilde{a}_{-k}$-pairs). Therefore, it should be related to the bare vacuum through the fermionic Bogoliubov transformation which mixes a_k with $\tilde{a}_k^\dagger$.

Denote the thermal vacua by $\langle 0(\theta)|$ and $|0(\theta)\rangle$, and their vacuum operators by $\alpha_k(\theta)$ and $\tilde{\alpha}_k(\theta)$. The vacuum of $(a_k, \tilde{a}_k)$ is denoted by $|0\rangle\rangle$. Thus we have

$$a_k|0\rangle\rangle = \tilde{a}_k|0\rangle\rangle = 0, \tag{3.39}$$

$$\alpha_k(\theta)|0(\theta)\rangle = \tilde{\alpha}_k(\theta)|0(\theta)\rangle = 0. \tag{3.40}$$

Obviously the vacuum operators are required to satisfy the fermionic anticommutation relations:

$$[\alpha_k(\theta), \alpha_l^\dagger(\theta)]_+ = \delta(\vec{k} - \vec{l}), \tag{3.41}$$

$$[\tilde{\alpha}_k(\theta), \tilde{\alpha}_l^\dagger(\theta)]_+ = \delta(\vec{k} - \vec{l}), \tag{3.42}$$

$$[\alpha_k(\theta), \tilde{\alpha}_l(\theta)]_+ = [\alpha_k(\theta), \tilde{\alpha}_l^\dagger(\theta)]_+ = 0. \tag{3.43}$$

As the relations (2.109) and (2.110) show, the thermal Bogoliubov transformation is

$$\begin{bmatrix} a_k \\ \tilde{a}_k^\dagger \end{bmatrix}^\mu = B_k^{-1}(\theta)^{\mu\nu} \begin{bmatrix} \alpha_k(\theta) \\ \tilde{\alpha}_k^\dagger(\theta) \end{bmatrix}^\nu, \tag{3.44}$$

$$[a_k^\dagger, \tilde{a}_k]^\mu = [\alpha_k^\dagger(\theta), \tilde{\alpha}_k(\theta)]^\nu B_k(\theta)^{\nu\mu} \tag{3.45}$$

with

$$B_k(\theta) = \begin{bmatrix} c_k & d_k \\ -d_k & c_k \end{bmatrix}. \tag{3.46}$$

with $c_k = \cos\theta_k$ and $d_k = \sin\theta_k$. The number density parameters n_k and $\tilde{n}_k$ are defined by $n_k[\delta(\vec{q})]_{\vec{q}=0} \equiv \langle 0|a_k^\dagger a_k|0\rangle$ and $\tilde{n}_k[\delta(\vec{q})]_{\vec{q}=0} \equiv \langle 0|\tilde{a}_k^\dagger \tilde{a}_k|0\rangle$. We then have

$$n_k = \tilde{n}_k = d_k^2. \tag{3.47}$$

This gives the parameter $d_k = \sin\theta_k$ in terms of the number parameter n_k.

This transformation is induced by $U_B(\theta)$ as

$$\alpha_k(\theta) = U_B(\theta)a_k U_B^{-1}(\theta), \tag{3.48}$$

$$\tilde{\alpha}_k(\theta) = U_B(\theta)\tilde{a}_k U_B^{-1}(\theta), \tag{3.49}$$

where $U_B(\theta)$ is

$$U_B(\theta) = \exp\left(-\int d^3k\, \theta_k[a_k\tilde{a}_k - \tilde{a}_k^\dagger a_k^\dagger]\right). \tag{3.50}$$

Then, we have $|0(\theta)\rangle = U_B(\theta)|0\rangle\rangle$, which gives

$$|0(\theta)\rangle = \exp\left([\delta(\vec{q})]_{\vec{q}=0}\int d^3k \ln\cos\theta_k\right)$$

$$\times \exp\left(\int d^3k \ln[1 + a_k^\dagger \tilde{a}_k^\dagger \tan\theta_k]\right)|0\rangle\rangle. \tag{3.51}$$

Presence of $[\delta(\vec{q})]_{\vec{q}=0}$ again makes amplitudes of all the finite number of thermal pair states vanish, indicating condensation of infinite number of thermal pairs. We thus have the parameterized Fock spaces $\mathcal{H}(\theta)$ which form an inequivalent set. Since the $\alpha(\theta)$-vacua are the states with a condensation of $a\tilde{a}$-pairs, they are invariant under tilde conjugation.

3.5 REFERENCES

[1] F. Bloch and A. Nordsieck, *Phys. Rev.* **52** (1937) 54.

[2] H. Umezawa, *Quantum Field Theory*, (North-Holland, Amsterdam, 1956).

[3] H. Umezawa and S. Kamefuchi, *Prog. Theor. Phys.* **6** (1951) 543.

4

Quasi-Particle Picture

4.1 Modern Particle Picture and Old Atomism

In the last chapter we saw several examples of continuously parameterized sets of inequivalent Fock spaces $\mathcal{H}(\theta)$ where two spaces with different θ are mutually inequivalent to each other. Here inequivalence means that a Fock space contains some vectors which are not a superposition of state vectors from another Fock space. This inequivalence is caused by a condensation of infinite numbers of particles into the vacuum. This is in sharp contrast to the situation in quantum mechanics. It is due to this fact that a system of quantum fields can manifest itself in a variety of phases, in each of which different kinds of observed particles describe the system of quantum fields. In this way quantum field theory is capable of dealing with many phases in a system.

Since a system is observed in many different guises, the modern particle picture has a dual structure in its language [1]: a language for fundamental entities and a language for the observed particles. The old atomism had a single structure: the fundamental particles were the observed ones. Since statistical mechanics is supposed to be the theory which derives macroscopic phenomena from the microscopic world, the change of structure of language for the microscopic world should influence the formulation of statistical mechanics. Modern statistical mechanics, therefore, needs certain significant modifications in order to accommodate the dual structure in its language.

Since appearance of inequivalent representations is of such a vital significance in formulating a modern many body theory, we present in this chapter a brief but systematic description of the mathematics of inequivalent representations in quantum field theory.

4.2 Countably Infinite Degrees of Freedom

As was shown in chapter 3, a quantum field can be expressed in terms of oscillator operators α_k satisfying

$$[\alpha_k, \alpha_l^\dagger] = \delta(\vec{k} - \vec{l}). \tag{4.1}$$

In applying the ideas in this section to the quantum fields studied in the last chapter, α_k stands either for a_k or for $\alpha_k(\theta)$ with any θ. The momentum suffix k in α_k shows that the quantum field system consists of an infinite number of oscillators; it has an infinite number of degrees of freedom. At first glance we may feel that we have a continuous set of canonical operators because $\vec{k}$ is a continuous variable. However the situation is more complicated. This is due to the fact that the wave function with a given momentum $\vec{k}$ is the plane wave $\exp[i\vec{k}\vec{x}]$, the norm of which is infinite. This is reflected in the presence of the δ-function in the above commutation relation. This commutation relation is meaningful only when it is smeared out over the space with a square integrable function, which describes a wave packet. This shows that the oscillator operators acting on vectors in $\mathcal{H}(\alpha)$ are not α_k but

$$\alpha_f = \int d^3k\, f(\vec{k})\alpha_k, \tag{4.2}$$

where $f(\vec{k})$ is a square integrable function:

$$\int d^3k |f(\vec{k})|^2 < \infty. \tag{4.3}$$

A well known example of an orthonormalized complete set of square integrable functions is given by the harmonic oscillator wave functions, which are classified by the principal quantum numbers. Any square integrable function can be expanded in terms of the oscillator wave functions. Thus any orthonormalized complete set of square integrable functions is a countably infinite set. We denote it by $\{f_i(\vec{k})\}$. We choose i as $i = 1, 2, 3 \cdots$. Then we have the countably infinite set of oscillator operators:

$$\alpha_i = \int d^3k f_i(\vec{k})\alpha_k. \tag{4.4}$$

This gives

$$[\alpha_i, \alpha_j] = \delta_{ij}. \tag{4.5}$$

As we saw in section 2.1, the $\alpha_k^\dagger$ increases the particle number defined by the eigenvalue of the operator $\alpha_k^\dagger \alpha_k$, and therefore, is called the creation operator. With the smeared-out process in (4.4) the operator $\alpha_i^\dagger$ creates a particle in a wave-packet state. We need to confine our attention to these wave-packet states, because the plane wave function is not normalizable (i.e. not square integrable). Although we need these wave-packet operators in creating particle states, we are going to find that *the basic operators used in algebraic manipulation are the plane wave operators* α_k. For example, the transformations (3.4), (2.109) and (2.110) were expressed, not in terms of the wave packet operators, but in terms of the plane wave operators α_k. This kind of formalism is possible because the square-integrable functions are Fourier transformable; any wave packet operators are given by a linear sum of plane wave operators.

In this section we saw that in quantum field theory we are dealing with a countably infinite degrees of freedom [1].

4.3 Continuous Set of Number States

In the last subsection it was shown that a quantum field system has an infinite number of degrees of freedom. There we introduced the wave packet oscillator operator α_i. Let us call the particle state created by $\alpha_i^\dagger$ the i-state. The particle number in the i-state is denoted by n_i. Thus each number sequence of $n_1 n_2 n_3 \cdots$ specifies a state with this set of particle numbers. This state, denoted by $|n_1 n_2 n_3 \cdots\rangle$, is called the number state. Any state of particles is a superposition of number states.

Let us first consider the subset $(|n_1 n_2 n_3 \cdots\rangle; n_i = 0$ or $1)$, which is called the fermionic subset. Now consider the set of numbers $0.n_1 n_2 n_3 \cdots$. This set covers the entire real number smaller than 1 in binary number system. Since real number is continuous, the fermionic subset is a continuous set; it has an uncountably infinite number of states.

We now consider the entire set of the number states. Since its subset is a continuous set, the entire set itself is also a continuous set [2].

However, since the probability for each state should be well defined, the state vectors should be normalizable, and therefore, any physically sensible state vector space has countable number of base vectors: it is separable. This means that the set of all of the number states is too large to be a base set for a physical state vector space. Let us denote the state with $n_i = 0$ for all i by $|0\rangle$ and call it the vacuum. We now consider the set $(|n_1 n_2 n_3 \cdots\rangle; \sum n_i = $ finite$)$. This is a countable set and it includes the vacuum. This set is called the 0-set [3, 4, 5, 1]. Since in reality we are interested in states of finite number of particles, even if that number is large, the state vector space built on the zero-set is a reasonable choice for our state vector space. This state vector space is called the Fock space.

However, this does not mean that the state vector space is unique, because there are an infinite number of choices of the 0-set for the following reason. As we saw in the last chapter, the definition of the vacuum $|0\rangle$ and the set of its annihilation operators α_k is not unique; for example, the vacuum $|0\rangle$ can be $|0\rangle\rangle$ with the annihilation operators a_k, or $|0(\theta)\rangle$ with $\alpha_k(\theta)$, and so on. There are many choices of vacuum depending on the structure of the condensate, each corresponding to its own choice of the 0-set.

These infinite number of choices for the vacuum originates from the fact that while the entire number states form a continuous set, the basis vectors of a Fock space form a countable set. There are an infinite ways of choosing a countable subset from a continuous set [3, 4, 5, 1].

4.4 Quasi-Particles and Dynamical Map

Since in quantum field theory we have an infinite choice of inequivalent representations which are classified by the structure of the particle condensations in the vacuum states, there is no guarantee that a given Lagrangian or Hamiltonian together with its canonical commutation relations uniquely specifies a physical solution; we should also specify the Fock representation associated with each given physical situation.

Consider the dynamics of electrons in a metal. They are usually described by normal electron spin up and down operators $a_{k,\uparrow}, a_{k,\downarrow}$. When this system appears to be a normal conductor, its vacuum $|0\rangle\rangle$ is empty of the normal electrons and its Fock space is built on the 0-set for $a_{k,\uparrow}, a_{k,\downarrow}$. However the same metal with the same Hamiltonian can appear also in the superconducting state. The superconducting vacuum $|0\rangle$ contains the condensation of Cooper pairs (each pair consisting of spin up and down electrons) and is related to the normal conducting vacuum through the fermionic Bogoliubov transformation, the generator of which is the same as the squeezed state transformation generator G_B in (3.19) with a_k and $\tilde{a}_k$ being replaced by the normal electron spin up and down operators, i.e. $a_{k,\uparrow}$ and $a_{k,\downarrow}$, respectively. Denoting the transformed operators by $\alpha_{k,\uparrow}$ and $\alpha_{k,\downarrow}$, the superconducting Fock space is built on the 0-set for these operators. The normal conducting and superconducting Fock spaces are inequivalent to each other. In a superconducting vacuum there are an infinite number of $(a_{k,\uparrow}, a_{k,\downarrow})$-pairs (called Cooper pairs) in the condensate.

In a similar manner, a system of quantum fields with magnetic moments may appear in a paramagnetic phase or in a ferromagnetic phase or many other magnetically ordered phases, depending on the way in which the magnetic spins condense in the vacuum. A system of quantum fields of atoms may appear in a gas phase, in a crystal phase or in many other kinds of phases.

These examples illustrate a general statement that the many different phases are manifestations of quantum field dynamics through different Fock spaces. They are distinguished from each other by the form of particle condensation in their vacuum which gives rise to different choices of 0-set. We will come back to this point in the next chapter where we explain the mechanism for the spontaneous breakdown of symmetries in which a variety of ordered states are created. Particle condensation in vacua plays an even wider role however. We will see in chapter 6 how particle condensation creates macroscopic objects. Further development of this line of approach leads us to a quantum field formalism for thermal phenomena.

When a Fock space is built on the 0-set for α_k, the particles created by $\alpha_k^\dagger$ are called the *quasi-particles*. A matrix element of an operator among any two vectors in this Fock space consists of two actions; one is annihilating some particles in the ket state and another is creating some particles in the bra state. Thus this operation is expressed by a product of the some creation operators $\alpha_k^\dagger$ and some annihilation operators α_k in such a manner that all the creation operators stand left to all the annihilation operators. Products of this form are called the *normal products*.

This consideration shows that the language of quantum field theory has a dual structure. The basic language consists of the Heisenberg operators, in terms of which the Lagrangian, Hamiltonian and canonical commutation relations for fields are expressed, while the phenomenological language consists of the quasi-particle operators. In order to put this aspect in a mathematical form, we introduce the concept of the *dynamical map* [6, 1]. Suppose we have a scalar Heisenberg field, $\psi(x)$, satisfying a certain Heisenberg equation. Assume that a physical situation of this system is described by state vectors in a Fock space $\mathcal{H}(\alpha)$ associated with the creation and annihilation operators $(\alpha_k, \alpha_k^\dagger)$ for free particles with energy ω_k. In other words, the Fock space $\mathcal{H}(\alpha)$ is built on the 0-set for α_k. This means that the action of any operator on state vectors in the Fock space $\mathcal{H}(\alpha)$ is defined by the expression of this operator in terms of normal products of the operators $(\alpha_k, \alpha_k^\dagger)$. We thus write a Heisenberg field operator as

$$\psi(x) = F[x : \alpha, \alpha^\dagger]. \tag{4.6}$$

This identifies the operation of the Heisenberg operator ψ acting on state vectors in $\mathcal{H}(\alpha)$. In other words, ψ acts as $F[x : \alpha, \alpha^\dagger]$ when the Fock space $\mathcal{H}(\alpha)$ is chosen.

Instead of using explicitly the creation and annihilation operators of the quasi-particle, we frequently use their free field $\varphi(x)$ which is called the quasi-particle field. Then we put (4.6) as

$$\psi(x) = F[x : \varphi]. \tag{4.7}$$

Both (4.6) and (4.7) are called the dynamical map of ψ [6, 1]. In a relativistic case $\varphi(x, t)$ has the form given in (1.43) with a being replaced by α:

$$\varphi(x) = \frac{1}{(2\pi)^{3/2}} \int d^3k \, \frac{1}{\sqrt{2\omega_k}} [\alpha_k e^{i(\vec{k} \cdot \vec{x} - \omega_k t)} + \alpha_k^\dagger e^{-i(\vec{k} \cdot \vec{x} - \omega_k t)}]. \tag{4.8}$$

In a nonrelativistic case it may have either the above form or

$$\varphi(x) = \frac{1}{(2\pi)^{3/2}} \int d^3k \, \alpha_k e^{i(\vec{k} \cdot \vec{x} - \omega_k t)}. \tag{4.9}$$

There are more complicated forms of φ such as the Dirac field and others. In this book φ stands for any quasi-particle field, while ψ does for any Heisenberg field.

The dynamical maps are true only for its matrix elements among state vectors in the Fock space $\mathcal{H}(\alpha)$ built on the 0-set for α_k. More precisely, the relation (4.6) means

$$\langle a|\psi(x)|b\rangle = \langle a|F[x : \alpha, \alpha^\dagger]|b\rangle, \tag{4.10}$$

where $\langle a|$ and $|b\rangle$ stands for any bra- and ket-vectors in $\mathcal{H}(\alpha)$. When a relation among operators holds true only for its matrix elements between vectors belonging to a particular state vector space, it is called a *weak relation*. Thus the dynamical maps are weak relations. To emphasize this point, we write (4.6) as

$$\psi(x) \overset{w}{=} F[x : \alpha, \alpha^\dagger]. \tag{4.11}$$

However, in the following, we mostly use the usual equality symbol unless we wish to emphasize the weak nature of a relation.

4.5   Time-Independent Global Operators

In (3.30) in section 3.3 we considered an operator which is a spatial integration of a local density:

$$G = \int d^3 x \rho(\vec{x}). \tag{4.12}$$

An operator of this form is called a global operator. In this section we study the dynamical map of time-independent global operators.

Let us first focus our attention on the term of the following form in the dynamical map of G:

$$\int d^3 k\, d^3 l\, d^3 p\; c(\vec{k}, \vec{l}, \vec{p}) \alpha_k^\dagger \alpha_l^\dagger \alpha_p \alpha_{k+l-p}. \tag{4.13}$$

Here $c(\vec{k}, \vec{l}, \vec{p})$ is a c-number function. The momentum is conserved because of the spatial integration in (4.12). The phase space which contributes to the matrix elements of this term shrinks because of the time independence which leads to the energy conservation

$$\omega_k + \omega_l = \omega_p + \omega_{k+l-p}, \tag{4.14}$$

where ω_k is the energy of the quasi-particle with momentum $\vec{k}$. Although we have nine integration variables, only eight variables are independent due to energy conservation. Thus matrix elements of this term between wave packet states vanish as $1/L$ with L being the linear size of the world which is infinite.

The situation is different for bilinear terms. Energy conservation forbids terms of the form $\alpha_k \alpha_l$ or $\alpha_k^\dagger \alpha_l^\dagger$ from appearing in this dynamical map because quasi-particle energies cannot be negative. Thus we have terms only of the form $\alpha_k^\dagger \alpha_l$. Momentum conservation requires $\vec{k} = \vec{l}$, leading to the form $\alpha_k^\dagger \alpha_k$. Note that this naturally conserves energy; there is no loss of phase space due to energy conservation. Thus matrix elements of this term survive. We thus see that there are no normal product terms of an order higher than second order in the dynamical map of a time independent global operator.

When there exists a quasi-particle χ which has zero energy at $\vec{k} = 0$:

$$\omega_0^\chi = 0, \tag{4.15}$$

certain terms linear in its creation or annihilation operators of the χ-field can survive. This can be shown as follows. Let χ_k denote the annihilation operator of this field. The spatial integration of the linear χ-term picks up the zero momentum contribution, giving rise to terms of the forms $\chi_k \delta(\vec{k})$ or $\chi_k^\dagger \delta(\vec{k})$. These terms naturally conserve energy due to the relation (4.15). Thus there is no loss of phase space due to energy conservation and these terms survive. When (4.15) holds, the particle is said to have gapless energy.

In summary we can say that the dynamical map of the time-independent global operator G has the structure

$$G = \sum_{ij} \int d^3 k [c_{ij}(k)\alpha_{i,k}^\dagger \alpha_{j,k} + (c\chi_k + c^* \chi_k^\dagger)\delta(\vec{k})]. \tag{4.16}$$

Here the c-number coefficients $c_{ij}(k)$ do not vanish only for those (i,j) with the same energy: $\omega_{i,k} = \omega_{j,k}$. The $\alpha_{i,k}$ is the annihilation operator of the i-th quasi-particle. The χ_k is the annihilation operator of a particle with gapless energy. If there is no gapless energy particle, G annihilates the vacuum ($G|0\rangle = 0$), because the $\chi^\dagger$ - term in (4.16) does not appear. Note that the relation (4.16) is a weak relation.

When a global operator does not annihilate the vacuum, i.e. $G|0\rangle \neq 0$, it has some pathological properties. First of all, the state $G|0\rangle$ is not normalizable when the vacuum is invariant under the time and space translations (the Coleman theorem). This can be seen as follows. The relation (4.12) gives

$$|G|0\rangle|^2 = \int d^3 x \int d^3 y \langle 0|\rho(\vec{x},t)\rho(\vec{y},t)|0\rangle. \tag{4.17}$$

The translational invariance of vacuum shows that

$$\langle 0|\rho(\vec{x},t)\rho(\vec{y},t)|0\rangle = F(\vec{x} - \vec{y}) \tag{4.18}$$

with a function F.

Exercise 1 Derive (4.18).

Hint: The Hamiltonian H and the momentum operator $\vec{P}$ are the generators for the time and space translation, respectively:

$$\rho(x) = e^{i[Ht - \vec{P}\cdot\vec{x}]}\rho(0)e^{-i[Ht - \vec{P}\cdot\vec{x}]}. \tag{4.19}$$

The translational invariance of the vacuum means $\exp[iHt]|0\rangle = |0\rangle$ and $\exp[i\vec{P}\cdot\vec{x}]|0\rangle = |0\rangle$, implying $\vec{P}|0\rangle = H|0\rangle = 0$.

We thus find

$$|G|0\rangle|^2 = \int d^3y F(\vec{y}) \int d^3x, \qquad (4.20)$$

which is infinity unless $G|0\rangle = 0$.

Furthermore, we have seen that when G is independent of time and does not annihilate the vacuum, there should be a χ-term in its dynamical map according to (4.16). Since this term carries a delta function, it is not well defined unless it is smeared out with a square integrable function $f(\vec{x})$. Thus such a G is an anomalous operator which can be defined with the smearing out trick given by

$$\int d^3x f(\vec{x})\rho(\vec{x}). \qquad (4.21)$$

The limit $f \to 1$ should be performed at the end of a calculation. In chapter 3 we have seen some examples of the anomalous operators, though they were not independent of time.

4.6 The Dynamical Map of the Hamiltonian

4.6.1 QUASI-PARTICLE FREE HAMILTONIAN

Since the Hamiltonian H is the generator for time translation, the fact that it commutes with itself implies that H is a time-independent global operator. Therefore, its dynamical map has the form in (4.16).

In this section we confine our attention to those vacua which are invariant under time translation. Then, $\exp[iHt]|0\rangle = |0\rangle$, which gives

$$H|0\rangle = 0. \qquad (4.22)$$

Therefore, the linear χ-terms in (4.16) do not appear. Diagonalizing the matrix $c_{ij}(k)$, we obtain

$$H = \sum_i \int d^3k \omega_k^i \alpha_{i,k}^\dagger \alpha_{i,k}. \qquad (4.23)$$

To simplify the notation, we consider only one field in the following. Thus we have

$$H = H_0 \qquad (4.24)$$

with

$$H_0 = \int d^3k\, \omega_k \alpha_k^\dagger \alpha_k. \qquad (4.25)$$

This is called the quasi-particle free Hamiltonian. The coefficient ω_k is the quasi-particle energy.

At first glance this result might surprise some readers. Therefore, we caution the readers that this relation is only a weak relation and it does not mean that the field has no interactions. It does not deny that field interactions can induce particle reactions; it implies only that particle reactions do not contribute to the energy when ω_k is chosen to be the renormalized one. In other words, although field interactions may influence the quasi-particle energy through their contribution to the self-energy, the total energy of the system is the linear sum of quasi-particle energies. This is the essential feature of the quasi-particle picture. To emphasize the weak nature of the relation, we may put the relation (4.24) in the form

$$H \stackrel{\text{w}}{=} H_0. \tag{4.26}$$

However we frequently use the usual equality symbol for a sake of simplicity.

To have a better understanding of this subject, we provide here another derivation of (4.26). For simplicity, we assume that the vacuum is also invariant under spatial translation. Then the dynamical map in (4.7) has the following explicit form:

$$\psi(x) = \sum_n \int d^4x_1 \, d^4x_2 \cdots c(x - x_1, \cdots, x - x_n) : \varphi(x_1)\varphi(x_2)\cdots\varphi(x_n) :$$

$$\tag{4.27}$$

where the notation $: A :$ means the normal product and φ stands for any quasi-particle field. It is important to note here that this is a weak relation so that we should consider its matrix elements among the state vectors in the space $\mathcal{H}(\alpha)$ and that the particle states are not plane waves but wave packets. Therefore, in these matrix elements all the oscillating functions are integrated with square integrable functions. Therefore there is no contribution from the boundary domain at $t = \pm\infty$. This means also that this dynamical map can be put in the Fourier form.

Now recall the Heisenberg equation

$$i\frac{\partial}{\partial t}\psi(x) = [\psi(x), H]. \tag{4.28}$$

On the left hand side the time derivative acts on the expansion coefficients $c(x - x_1 \cdots)$. Since there is no contribution from the boundary surface when integration by parts is utilized, the time-derivative acts on each φ. Use of the relation

$$i\frac{\partial}{\partial t}\varphi(x) = [\varphi(x), H_0] \tag{4.29}$$

gives

$$\langle a|[\psi, H]|b\rangle = \langle a|[\psi, H_0]|b\rangle \tag{4.30}$$

for any operator ψ and any two vectors, $|a\rangle$ and $|b\rangle$, in the Fock space $\mathcal{H}(\alpha)$. This shows

$$H \stackrel{\text{w}}{=} H_0 + c\text{-number}. \tag{4.31}$$

By adjusting the c-number constant in H in such a way that $H|0\rangle = 0$, we obtain the relation (4.26).

4.6.2   REACTIONS AMONG QUASI-PARTICLES

In the last subsection we saw that the Hamiltonian is weakly equal to the free Hamiltonian. This leads us to the question of how particle reactions can take place when the total energy is the sum of free particle energies. The answer can be found in the study of the dynamical map of the time-independent global operator in section 4.5. There it was shown that the contribution to the matrix elements of G from normal product terms of order higher than second order in the dynamical map of a time-independent global operator G is of order of $(1/L)$ with L being the linear size of the system. At the end of the calculation of matrix elements the limit $L \to \infty$ should be performed. Applying this argument to H, we may write

$$H = H_0 + Q, \tag{4.32}$$

where Q stands for the higher order normal products with the matrix elements of order of $1/L$. In calculation of the total energy, the limit $L \to \infty$ eliminates the contribution of Q. However, Q does contribute to the S-matrix. When Q is treated as a perturbation interaction, the contribution of Q to the S-matrix is integrated over the reaction time which is order of L. Thus this contribution survives even at the limit of infinite L. A straightforward derivation of the dynamical map, (4.26), of the Hamiltonian from a perturbation calculation and a consideration of this structure of the S-matrix were presented in [1].

The dynamical map, (4.26), of the Hamiltonian was explicitly obtained in some exactly solvable models such as the $N\theta$- and Lee models [7, 1].

4.6.3   ASYMPTOTIC FIELDS AND QUASI-PARTICLES

In the quasi-particle picture dynamical maps of all the operators are expressed in terms of free quasi-particle operators. This leads us to the question of what kind of particles these quasi-particles are. In this subsection we analyze this question.

To make the question intuitive let us consider some reactions among particles. There a free particle picture naturally shows up. The particle reactions are transitions between the initial state of free incoming particles and the final state of free outgoing particles. This free particle picture originates from experiments which identify the incoming free particles to specify the initial state and the outgoing free particles for the final states. This suggests the possibility that these asymptotic particles, either the incoming or outgoing particles, may be usable as the quasi-particles.

However, this does not settle the question. We may say that the incoming particles are free because the experimental setup prepares the spatially separate incoming beams. However, it is not obvious how the outgoing particles naturally become free. Before 1950 physicists enforced this picture by introducing the so called adiabatic trick in which coupling constants, say

g, are replaced by $\exp\left[-\epsilon|t|\right]g$; the limit of vanishing ϵ was to be performed at the end of calculations. This method is artificial, though it is frequently useful in perturbation calculation. It can be justified only when we know that the interaction without the adiabatic factor naturally disappears at the asymptotic domain $t = \pm\infty$.

The reason for disappearance of mutual interactions among particles comes from the wave packet picture. We have seen in section 4.2 that particle states should be wave packet states because the plane wave states are not normalizable. On the other hand wave packets spread in time. Denoting the size of wave function at time t by $v(t)$, this tends to infinity after an infinite time. The probability of finding another particle inside of this wave function is $(1/v(t))$. Thus the coincidence probability at a specific point for two particles is $(1/v(t))^2$. However, since the coincidence point can be any point we should multiply $v(t)$ with this in order to obtain the overlapping probability of two wave functions. Thus the overlapping probability of two wave functions diminishes in time as $(1/v(t))$. In this way the spread of the wave function eliminates the mutual interaction effects. When the wave front of the wave function spreads with a constant velocity (for example the velocity of light in relativistic models), the radius increases linearly in time so that $v(t)$ is proportional to t^3. Then the overlapping probability decreases in time as t^{-3} so that the interaction effect in the S-matrix diminishes as $t^{-3/2}$.

Exercise 2 Consider a particle A which decays to two particles as $A \rightarrow B + C$. There is the regeneration process; the two decay products, B and C, meet again and produce the parent particle A. This slightly modifies the simple exponential law for probability of finding the decaying A-particle. Using the above picture of behavior of the wave packet, show that one regenerating process creates the regenerated A-wave function with the amplitude proportional to $g^2 t^{-3/2}$. Here g is the decay coupling constant.

Comment: This effect was first discovered by J. Schwinger in [8]. According to this paper, the intuitive understanding of this effect in terms of spreading wave packet was given by Haag.

This analysis of asymptotic behavior of the mutual interactions was developed by Haag [9]. Lehmann, Symmanzik and Zimmermann formulated this idea in terms of asymptotic limit of quantum fields [10]. The main idea is the following. Consider a wave packet function for a particle with quasi-particle energy ω_k:

$$f(x) = \int \frac{d^3k}{(2\pi)^{3/2}} f(\vec{k}) e^{-i[\vec{k}\cdot\vec{x}-\omega_k t]}. \tag{4.33}$$

Let us begin with the quasi-particle field φ of the nonrelativistic form in (4.9). Consider the matrix elements

$$F_{ab}(t) \equiv \langle a| \int d^3x\, f(x)\psi(x)|b\rangle \tag{4.34}$$

with two states belonging to the Fock space $\mathcal{H}(\alpha)$. Note that $f(x)$ is a single particle wave function. It is due to the Riemann-Lebesgue theorem that the $t \to \pm\infty$ limiting process applied to F_{ab} picks up the part of $\psi(x)$ which oscillates with the single particle frequency ω_k. Thus this limit gives rise to the matrix elements which are the same as those of the quasi-particle field. This defines the quasi-particle annihilation operators α_k as

$$\lim_{t \to \pm\infty} F_{ab}(t) = \langle a| \int d^3k f(\vec{k}) Z^{1/2}(\vec{k}) \alpha_k |b\rangle. \tag{4.35}$$

with a wave function renormalization factor $Z(\vec{k})$.

In a relativistic model φ has the form (4.8). Since this φ contains both negative and positive frequency parts, we need a more complicated process in order to separate them from each other. To find this process we note the relation

$$\int d^3x f(x) \overset{\leftrightarrow}{\partial_t} \varphi(x) = -i \int d^3k \sqrt{2\omega_k} f(\vec{k}) \alpha_k, \tag{4.36}$$

where use was made of the notation $\overset{\leftrightarrow}{\partial} = \overset{\rightarrow}{\partial} - \overset{\leftarrow}{\partial}$. This relation shows that we should replace the above F_{ab} by

$$F_{ab}(t) \equiv \langle a| \int d^3x f(x) \overset{\leftrightarrow}{\partial_t} \psi(x) |b\rangle, \tag{4.37}$$

This gives

$$\lim_{t \to \pm\infty} F_{ab}(t) = Z^{1/2} \langle a| \int d^3k \sqrt{2\omega_k} f(\vec{k}) \alpha_k |b\rangle. \tag{4.38}$$

In this case Z becomes independent of $\vec{k}$ due to the relativistic invariance of the model.

Exercise 3 Derive the relation (4.36) by using (4.8).

The limit with $t = -\infty$ defines the incoming quasi-particle field, while the one with $t = \infty$ gives the outgoing quasi-particle field. Since this time limit is a limiting process applied to the matrix elements associated with the particular Fock space, it is a weak limit. Since the outgoing particle states belong to the Fock space of the incoming particles, we can choose either the incoming or outgoing fields for the quasi-particle field φ.

Dynamical maps written in terms of the asymptotic free fields are called Haag expansions [9]. Using these limiting processes Lehmann et al derived a general expression for the Haag expansion [10].

4.6.4   MANY CHOICES FOR INCOMING PARTICLES

The consideration of the last subsection might suggest that the choice of incoming (or outgoing) particles is unique because it is the asymptotic limit of the Heisenberg fields. This idea is wrong because there are many choices for incoming particles; a system in a superconducting state requires its own choice for the incoming electrons, while the system in a normal conducting state has a different choice of incoming electrons.

This is allowed because the time limit used in the last section is a weak one. Suppose that we wish to use the incoming free particles for the quasi particles. Since the limit is the weak one, we should know how the Fock space is associated with the incoming particles before we perform the limiting process. Then, the time limit should reproduce the same quasi-particle. In other words we should prepare the annihilation and creation operators $(\alpha_k, \alpha_k^\dagger)$, out of which the Fock space $\mathcal{H}(\alpha)$ is made, in such a manner that the above limiting process should reproduce the same α_k. This is a self-consistent procedure. Since there exist an infinite number of inequivalent Fock spaces in quantum field theory, a particular Fock space should be chosen in a self-consistent manner.

A system manifests itself in a varieties of phases, and each phase uses its own choice of quasi-particles, which can be incoming or outgoing particles associated with this phase. The quasi-particles are chosen in a self consistent manner.

4.6.5   RENORMALIZATION

With the study of inequivalent Fock spaces developed in this chapter we now understand that renormalization is a part of the self-consistent method mentioned above. For example, we should prepare the quasi-particle energy ω_k in a self-consistent manner. Indeed, the time limiting process discussed in the last subsection cannot pick out the quasi-particle operators unless we prepare the right choice for ω_k. This argument can be supported by the following intuitive argument. An interaction among quantum fields induces, not only the particle reactions, but also a shift of particle energies by creating self-energies. Since the spread of wave packets eliminates only the mutual interactions, the interaction effect causing the self energies do not disappear at the asymptotic limit. This is the reason why we should prepare the unperturbed Hamiltonian so that it includes the self-energies, creating energy counter terms in the interaction Hamiltonians. In other words, the interaction Hamiltonian including the self-energy counter terms takes care of the mutual interactions among quasi particles. This is the reason for the energy renormalization (mass renormalization in relativistic cases).

The renormalization is, therefore, required for the sake of self-consistency in choice of Fock space. This is why the renormalization is made even in condensed matter physics where there are no ultraviolet divergence problems. The self-consistent renormalization gives rise to equations which determine energies of quasi-particles. These quasi-particle energy equations are different in Fock spaces for different phases; the energy equation for electrons in superconducting phase is different from the one for normal conducting phase.

This self-consistent view for renormalization plays a vital role in understanding the phenomena of spontaneous breakdown of symmetries and appearance of many kinds of ordered states. This is the main subject of the next chapter.

4.7 REFERENCES

[1] H. Umezawa, H. Matsumoto, and M. Tachiki, *Thermo Field Dynamics and Condensed States*, (North-Holland, Amsterdam, 1982).

[2] K. O. Friedrichs, *Mathematical Aspects of the Quantum Theory of Fields*, (Interscience, New York, 1953).

[3] L. Gärding and A. S. Wightman, *Proc. Nat. Acad. Sci. USA* **40** (1954) 617.

[4] L. Gärding and A. S. Wightman, *Proc. Nat. Acad. Sci. USA* **40** (1954) 622.

[5] A. S. Wightman and S. S. Schweber, *Phys. Rev.* **98** (1955) 812.

[6] L. Leplae, R. N. Sen and H. Umezawa, *Suppl. Prog. Theor. Phys.*, Communication issue for the 30th Anniversary of the Meson Theory by Dr. H. Yukawa, (1965) 637.

[7] J. Rest, V. Srinivasen, and H. Umezawa, *Phys. Rev.* **D3** (1971) 1890.

[8] J. Schwinger, *Ann. of Phys.* **9** (1960) 169.

[9] R. Haag, *Phys. Rev.* **112** (1958) 669.

[10] H. Lehmann, K. Symanzik, and W. Zimmermann, *Nuovo Cim.* **6** (1957) 319.

5

Ordered States

5.1 Spontaneously Broken Symmetries

5.1.1 A Physical Picture

A fascinating and significant phenomenon caused by particle condensation in vacua is the spontaneous breakdown of symmetries. Symmetry and order are two concepts which are mutually complementary. When we have rotational symmetry, there is no particular direction singled out as being different. On the other hand, a state in which a specific direction is noticeable is a state with directional order. To create such an ordered state, the rotational symmetry should be lost.

A fascinating situation is that although the Lagrangian and Hamiltonian are rotationally invariant, a vacuum can still manifest a directional ordered state. Such an ordered state can be created, for example, when particles with spin up condense into the vacuum, creating a macroscopic spin moment. Parallel and perpendicular directions with respect to the spin have different properties. This phenomenon is that of spontaneously broken rotational symmetry or the creation of a spin directional ordered state. In this situation, we may intuitively expect two features.

One is a result of original rotational invariance of the dynamics, which implies that the spin-up direction can be any direction. In other words all the states with different directions of the macroscopic spin must have the same energy, forming a continuous set of vacua with the same energy. The presence of this *continuous set of degenerate vacua*, $|0(\theta)\rangle$, is the first feature. Here θ is the parameter specifying the direction of the macroscopic moment. A vacuum of this kind can be created, for example, by condensation of particles with up-spin.

Another feature is a result of the stability of such an ordered state. Suppose that a spin in this particle condensate happens to deviate from the direction of the macroscopic moment. In order for the ordered state to be stable, this deviation of spin direction should be detected by other particles which would then have some means of correcting the wrong direction of the particular spin. This requires the presence of some communication device among the particles in the condensate; with this communication device any deviation of local spin from the direction of the macroscopic moment can be corrected by the system. This device is a wave which can propagate over the whole of the domain. In quantum field theory this range is usually infinite. Considering the statement of quantum theory about particle and wave, this implies the existence of certain quantum particles associated with the wave. The infinite range of the wave propagation means that the smallest value of energy of this particle is zero, meaning that the particle carries a gapless energy (in relativistic models, massless energy).

A well known example of this kind is a ferromagnet. It is obvious that the ferromagnets with all of possible angles of magnetic moment form a continuous set of energy degenerate states, and it is known that in this system there is the gapless energy particle called the spin wave or magnon. The presence of degenerate vacua and gapless energy particles are common features of states with spontaneously broken symmetry of any kind.

5.1.2   A MATHEMATICAL PICTURE

The intuitive picture presented in the last subsection can be put into a mathematical form which follows immediately from the analysis of the structure of time independent global operators presented in section 4.5.

Suppose that a Lagrangian for a Heisenberg field $\psi(x)$ is invariant under a continuous transformation

$$\psi(x) \rightarrow \psi'(x) = \psi(x, \theta), \tag{5.1}$$

with a continuous parameter θ. Let us denote the generator of this transformation by N:

$$\psi'(x) = e^{i\theta N} \psi(x) e^{-i\theta N}. \tag{5.2}$$

Let us assume that N is a global operator; that is, it is a spatial integration of a local operator of the form (4.12). In a later section we will show how to construct this global operator associated with each transformation. The invariance of the Lagrangian leads to the invariance of the Hamiltonian: $H = \exp[iN\theta] H \exp[-iN\theta]$. Thus we have

$$[H, N] = 0. \tag{5.3}$$

This implies that N is independent of time. Thus, N is a time independent global operator. This is a story about invariance, and the way in which this invariance manifests in phenomena depends on the choice of vacuum. The form of manifestation of the invariance is called the symmetry.

Let $|0\rangle\rangle$ denote a vacuum which is invariant under this transformation:

$$e^{i\theta N} |0\rangle\rangle = |0\rangle\rangle. \tag{5.4}$$

This is equivalent to the condition

$$N|0\rangle\rangle = 0. \tag{5.5}$$

Suppose that a form of condensation of particles with a finite density creates a new vacuum, $|0\rangle$, which is not invariant under the N-transformation:

$$N|0\rangle \neq 0, \tag{5.6}$$

This means that the N-symmetry is spontaneously broken in the Fock space built on this vacuum. The two conditions, (5.3) and (5.6), define the phenomenon of spontaneously broken symmetry.

Since N is a time independent global operator, its dynamical map has the form given in (4.16):

$$N = \int d^3 k [c_{ij}(k)\alpha^\dagger_{i,k}\alpha_{j,k} + (c\chi_k + c^*\chi^\dagger_k)\delta(\vec{k})]. \qquad (5.7)$$

Here the $c_{ij}(k)$ do not vanish only for those (i,j) with the same energy: $\omega_{i,k} = \omega_{j,k}$. $\alpha_{i,k}$ is the annihilation operator of the i-th quasi-particle while χ_k is the annihilation operator of a particle with gapless energy. If there were no gapless energy particle, N would annihilate the vacuum, because the $\chi^\dagger$ - term in (5.7) would not appear. Therefore, a non-invariant vacuum requires the existence of the gapless energy boson whose free field is denoted by $\chi(x)$. This gives us the Goldstone theorem which states that in any state with a spontaneously broken symmetry there exist certain energy gapless fields χ. This theorem had been illustrated by many examples in condensed matter physics such as phonons in crystals, magnons in ferromagnets, etc. Around 1961 Nambu, Goldstone and other physicists grasped this as a general theorem [1, 2]. Thus this is called the Nambu - Goldstone theorem and the χ-particles are called the Nambu - Goldstone (NG) particles. The first proof of this theorem was given in the article [3]. The proof has been further elaborated by the hands of many physicists. In this subsection we used a different form for the proof, because this proof not only shows the existence of NG particles but also indicates a general rule for the form of symmetry breakdown, leading us to the notion of symmetry rearrangement. This point will be clarified in the next subsection.

According to the consideration in 4.5, a time independent global operator which does not annihilate the vacuum is an anomalous operator. Therefore, N is an anomalous operator; its action on vectors in the Fock space built on the vacuum $|0\rangle$ does not belong to the same Fock space. Thus, the vectors $|0(\theta)\rangle = \exp[i\theta N]|0\rangle$ can form a continuous set of vacua. Since N commutes with the Hamiltonian, it does not induce any change in the vacuum energy. Hence this set is a continuous set of energy degenerate vacua on each member of which a Fock space is built. We thus have the continuous set of Fock space $\{\mathcal{H}(\theta)\}$. This situation is the same as the one discussed in chapter 3, where the coherent or squeezed states were studied.

In this way the intuitive argument of the last subsection is reproduced in a mathematical form. The only touchy point is that, since the operator used, N, is an anomalous operator, it needs a formal mathematical trick such as the smearing out trick. In a later section, we provide another mathematical proof of the Nambu - Goldstone theorem by means of the Ward - Takahashi relation [4, 5]. However, use of N provides us with a simple way of understanding the subject of spontaneously broken symmetries. We therefore continue to use N in the next subsection.

5.1.3 SYMMETRY REARRANGEMENT

When the transformation $\exp[i\theta N]$ is performed, the first term on the right hand side of (5.7) generates a mixing between quasi-particles with the same energy. Thus, these particles form a representation of the observed symmetry which is different from the original N-symmetry. The second term in (5.7) generates the translation of the Nambu - Goldstone field χ:

$$\chi(x) \to \chi(x) - \theta c^*, \tag{5.8}$$

as was shown in chapter 3 for the case of the coherent states. Thus, the N-transformation appears in observations as a mixing of energy-degenerate quasi-particles and as a translation of the Nambu - Goldstone field. We now have a general theorem: *When there is a spontaneous breakdown of symmetry, there should exist gapless energy fields and the symmetry transformation is induced by the boson translation of these gapless energy boson fields together with a possible mixing of energy degenerate quasi-particle fields.*

In this way the phenomenon of spontaneously broken symmetries can be understood as a change of form of the symmetry in going from the Heisenberg field level to the observed level. This is called the *symmetry rearrangement* [6, 7, 8]. Since the broken symmetry tends to create differences among the energy spectra of quasi-particles and since energy differences prohibit particle mixing, many singlet fields usually appear, which do not change under the N - transformation. This is called the freezing phenomenon.

In general terms the symmetry rearrangement is defined in the following way. Suppose that the Lagrangian, and therefore, the Hamiltonian are invariant under a continuous transformation with a continuous parameter θ:

$$\psi \to \psi' = Q[\psi : \theta]. \tag{5.9}$$

Now write the dynamical map of the Heisenberg field ψ in terms of the quasi-particle field φ as

$$\psi(x) = \psi(x : \varphi). \tag{5.10}$$

Assume that a continuous transformation

$$\varphi \to \varphi' = q[\varphi : \theta] \tag{5.11}$$

induces the Q-transformation as

$$Q[\psi(x) : \theta] = \psi(x : q[\varphi(x) : \theta]). \tag{5.12}$$

Then we state that the Q-symmetry is rearranged to q-symmetry.

In this discussion of spontaneously broken symmetry, the N-symmetry is rearranged to give translation of the gapless energy field χ together with possible mixing of energy degenerate quasi-particles. The essential part is the translation of χ-field; mixing of particles does not always happen because of the freezing mechanism.

In chapter 3 we saw that a boson field translation causes boson condensation. Therefore, we see that what creates the continuous set of degenerate vacua is the χ-translation which regulates the condensation of the NG particles.

In the following sections we present some examples of ordered states and symmetry rearrangement.

5.1.4 BOSON TRANSFORMATION

Consider a bosonic quasi-particle field φ. Since this is a free field, it satisfies a linear homogeneous differential equation:

$$\Lambda(\partial)\varphi(x) = 0 \tag{5.13}$$

and the bosonic commutation relation. This equation is invariant under the transformation

$$\varphi(x) \rightarrow \varphi(x) + f(x), \tag{5.14}$$

where the c-number function $f(x)$ satisfies the same free field equation:

$$\Lambda(\partial)f(x) = 0. \tag{5.15}$$

This is called the boson transformation. The bosonic commutation relation is also invariant under this boson transformation. Recalling the coherent state discussed in chapter 3, we see that the boson transformation induces a condensation of quasi-particles. We will see in the next chapter that the boson transformation plays a crucial role in the creation of macroscopic objects in quantum field theory.

Since a χ-particle has no energy gap (i.e. $\omega_k = 0$ for $\vec{k} = 0$), its free field equation

$$\Lambda_\chi(\partial)\chi(x) = 0 \tag{5.16}$$

has the property

$$\Lambda_\chi(0) = 0. \tag{5.17}$$

Therefore any constant c-number satisfies the free field equation for χ. Thus, the boson translation in (5.8) with the constant c is a particular case of the boson transformation (5.14). We find therefore that the symmetry rearrangement mentioned above carries the boson transformation of the NG boson. This is the prelude to the story of topological objects in ordered states. As we will learn in the next chapter, the origin of many defects in ordered states is due to the condensation of the NG bosons induced by a boson transformation.

5.1.5   LOW ENERGY THEOREM I

The fact that the invariant N-transformation always carries the translation of the Nambu - Goldstone boson implies the following theorem: *The
dynamical map of any N-invariant operators never contains χ without
derivatives.* Let I_N stands for operators which are invariant under the N-
transformation:

$$[I_N, N] = 0. \tag{5.18}$$

We denote by ϕ all the quasi particle fields except the NG fields. Then, the
above theorem is expressed by the following form of the dynamical map of
I_N:

$$I_N = I[\phi, \partial\chi]. \tag{5.19}$$

Thus, soft GN bosons do not contribute to the matrix elements of invariant
operators. This theorem is called the low energy theorem [8].

In quantum field theory particle reactions are described as a transition
from the incoming quasi-particle fields to the outgoing fields. Since this
point has been explained in many text books, we do not explain this here.
It is pointed out that the in-fields and out-fields are related to each other
through the operator called the S-matrix. The S-matrix is invariant under
all the invariant transformations. Therefore the S-matrix is invariant under
the N-transformation. Applying the above low energy theorem to the S-
matrix, we find:

$$S = S[\phi, \partial\chi]. \tag{5.20}$$

This indicates that *soft NG particles do not participate in any reactions*
[8]. When this theorem is applied to the chiral symmetry theory for pion
physics, in which the pion is considered as the Nambu - Goldstone particle,
it results in the Adler theorem which states that the zero energy pion
does not contribute to any reactions [9]. When this low energy theorem is
applied to the magnons, it agrees with Dyson's result which showed that
the S-matrix elements associated with lower energy magnons are smaller
[10].

5.1.6   GENERATORS AND THE NÖTHER CURRENT

In this section we discuss the problem of how we find the generator N
for a given transformation when a Lagrangian is given. Note that in this
section we do not assume the invariance of the Lagrangian under the
N-transformation. The structure of N depends on the choice of the Lagrangian, because the canonical conjugates of fields are determined by the
Lagrangian.

Consider a continuous transformation and write its infinitesimal one as

$$\psi(x) \rightarrow \psi'(x) = \psi(x) + \theta\delta\psi(x) \tag{5.21}$$

with the infinitesimal parameter θ. For simplicity, we considered only those transformations in which the space-time coordinate x does not change. A general argument including a change of x can be found in many text books on quantum field theory (see, for example, page 227 of [8]). The change of the Lagrangian density is denoted by $\theta\delta\mathcal{L}(x)$:

$$\theta\delta\mathcal{L}(x) = \mathcal{L}[\psi(x)] - \mathcal{L}[\psi'(x)]. \tag{5.22}$$

Exercise 1 Derive the relation

$$\delta\mathcal{L}(x) = \partial^\mu N_\mu, \tag{5.23}$$

where

$$N_\mu = \frac{\partial\mathcal{L}}{\partial\psi^\mu}\delta\psi. \tag{5.24}$$

Here $\psi_\mu = \partial_\mu\psi$.

Hint: Make use of the Euler equation given by the Lagrangian.

The relation (5.23) is called the Nöther theorem and the N_μ is called the Nöther current [11]. When we denote the canonical conjugate of ψ by π_ψ, we have $\pi_\psi = \partial\mathcal{L}/\partial\dot{\psi}$. Thus,

$$N_0(x) = \pi_\psi(x)\delta\psi(x). \tag{5.25}$$

Now define

$$N(t) = \int d^3x N_0(x) \tag{5.26}$$

and assume that $\delta\psi$ does not contain π_ψ when it is expressed in terms of ψ and π_ψ. Then, (5.26) gives

$$[\psi(x), N(t)]_{t_x=t} = i\delta\psi(x). \tag{5.27}$$

This shows that $N(t)$ is the generator for the transformation $\psi \to \psi'$. This is how the generator is constructed for a given transformation. Note that N *is a global operator*. The Nöther theorem (5.23) shows that

$$\dot{N} = \int d^3x\delta\mathcal{L}(x). \tag{5.28}$$

5.1.7   THE WARD - TAKAHASHI RELATIONS

Exercise 2 Prove the relation

$$\frac{\partial}{\partial t_x}\langle 0|T[A(x), B(y)]|0\rangle = \delta(t_x - t_y)\langle 0|[A(x)B(y)]|0\rangle$$
$$+ \langle 0|T[\dot{A}(x)B(y)]|0\rangle. \tag{5.29}$$

Hint: Use the relations

$$\langle 0|T[A(x)B(y)]|0\rangle = \theta(t_x - t_y)\langle 0|A(x)B(y)|0\rangle$$
$$+\theta(t_y - t_x)\langle 0|B(y)A(x)|0\rangle. \qquad (5.30)$$

and

$$\frac{\partial}{\partial t_x}\theta(t_x - t_y) = \delta(t_x - t_y). \qquad (5.31)$$

Repeating this process we can prove

$$\frac{\partial}{\partial t} \quad \langle 0| \quad T[N(t), \psi(x_1)\cdots\psi(x_n)]|0\rangle$$

$$= \sum_{a=1}^{n} \delta(t - t_a)\langle 0|T[\psi(x_1)\cdots[N(t), \psi(x_a)]\cdots\psi(x_n)]|0\rangle$$

$$+\langle 0|T[\dot{N}(t)\psi(x_1)\cdots\psi(x_n)]|0\rangle. \qquad (5.32)$$

Considering the relations (5.27) and (5.28), we obtain

$$\frac{\partial}{\partial t} \quad \langle 0| \quad T[N(t)\psi(x_1)\cdots\psi(x_n)]|0\rangle$$

$$= -i\sum_{a=1}^{n} \delta(t - t_a)\langle 0|T[\psi(x_1)\cdots\delta\psi(x_a)\cdots\psi(x_n)]|0\rangle$$

$$+\int d^3x\langle 0|T[\delta\mathcal{L}(x)\psi(x_1)\cdots\psi(x_n)]|0\rangle. \qquad (5.33)$$

Integrate both sides over the entire time domain. Assuming that no particles with infinite propagation range exist, we drop the effects from $t = \pm\infty$. We then obtain

$$\sum_a \quad i\langle 0|T[\psi(x_1)\cdots\delta\psi(x_a)\cdots\psi(x_n)|0\rangle$$

$$= \int d^4x\langle 0|T[\delta\mathcal{L}(x)\psi(x_1)\cdots\psi(x_n)]|0\rangle. \qquad (5.34)$$

These are called the Ward - Takahashi relations which are frequently useful for model-independent analysis of symmetries.

5.1.8 θ-SELECTION

According to (5.28) N appears to be a time independent global operator when the Lagrangian is invariant under the N-transformation. Thus, when the N-symmetry is spontaneously broken, i.e.

$$N|0\rangle \neq 0, \qquad (5.35)$$

N is an anomalous operator. In this case the derivation of (5.28) becomes tricky. In derivation of (5.28) we ignored the spatial integration of $\vec{\nabla}\vec{N}$.

However, when the N-symmetry is spontaneously broken, the Nambu - Goldstone field χ appears and its effect has an infinite range. It is then not obvious if the spatial integration of ∇N really vanishes. To avoid this difficulty we make use of the following trick. Introduce an operator $\Phi(x)$ which is not invariant under the N-transformation:

$$\delta\Phi(x) = i[N, \Phi(x)] \neq 0. \tag{5.36}$$

We then add $\epsilon\Phi(x)$ to the N-invariant Lagrangian density with a very small ϵ. Thus, the Lagrangian has the structure

$$\mathcal{L} = \mathcal{L}_N + \epsilon\Phi(x), \tag{5.37}$$

where $\mathcal{L}_N$ is the original N-invariant Lagrangian. The symmetry breaking term $\epsilon\Phi$ selects a particular way of symmetry breaking. (For example, in a ferromagnet, it corresponds to a choice of direction for the magnetic moment.) Now we have

$$\delta\mathcal{L} = \epsilon\delta\Phi. \tag{5.38}$$

As long as ϵ is finite, the χ field is not of gapless energy and we can safely ignore the spatial integration of ∇N. At the end of calculations the limit $\epsilon \to 0$ should be performed in order to recover the N-invariance. At this limit a particular value of the parameter θ is selected as it can be seen from the equation (5.34). Thus, the $\epsilon\Phi$-term is called the θ-selecting term.

To understand the role of the θ-selecting term, we suppose spontaneous break down of a spin-rotational symmetry. Then the parameter θ specifies a spin direction. Denoting the spin density operator by s_i with $i = 1, 2, 3$, we use ϵs_3 for the θ-selecting term. Then the total Lagrangian is invariant under the rotation around the third axis. This creates a system with macroscopic spin moment in the third direction, and in the limit of vanishing ϵ leaves the vacuum with macroscopic spin in the third direction. Thus among all of the degenerate vacua $|0(\theta)\rangle$ covering all the possible directions this particular direction is selected even at the limit of vanishing ϵ. In this case, $\langle 0|s_3|0\rangle$ is the macroscopic spin moment. The change δS_3 generated by S_1 or S_2 is respectively given by S_2 or $-S_1$. The double change $\delta\delta S_3$ due to doubly repeated use of S_1 or S_2 comes back to $(-S_3)$ and its vacuum expectation value produces the macroscopic spin moment with minus sign.

Generalizing this situation, we choose Φ in such a manner that we have

$$v \equiv -\langle 0|\delta\delta\Phi(x)|0\rangle \neq 0. \tag{5.39}$$

even at the limit $\epsilon \to 0$. This quantity v is called the order parameter, because it specifies the particular order selected by the θ-selecting term.

Now use of the Ward - Takahashi relation (5.34) together with (5.38) and ψ being replaced by $\delta\Phi$ gives

$$iv = -\epsilon \int d^4x \langle 0|T[\delta\Phi(x)\delta\Phi(y)]|0\rangle. \tag{5.40}$$

Note that this does not depend on t_y because integrand on the right hand side is a function of $(t_x - t_y)$.

It is known that any two point causal function has the Fourier form:

$$\langle 0|T[A(x)B(y)]|0\rangle = \int d^4k e^{ik^\mu (x-y)_\mu} \int d\kappa \frac{\rho(\vec{k}, \kappa)}{k_0^2 - \kappa^2 + i\epsilon_F}, \qquad (5.41)$$

or

$$\langle 0|T[A(x)B(y)]|0\rangle = \int d^4k e^{ik^\mu (x-y)_\mu} \int d\kappa \frac{\rho(\vec{k}, \kappa)}{k_0 - \kappa + i\epsilon_F}, \qquad (5.42)$$

with the Feynman infinitesimal ϵ_F. (The choice among these two expressions depend on the quasi-particle fields; either with only the positive frequency part or with both negative and positive frequency parts.) When this expression is applied to (5.40), we find that $\rho(0, \kappa)$ should not vanish at $\kappa^2 = Z\epsilon$ or $\kappa = Z\epsilon$ (depending on which of the above two relations is used) with a constant Z at the vanishing limit of ϵ, because otherwise v would vanish. This shows that the dynamical map of $\delta\Phi$ contains a linear term whose energy is gapless, implying the existence of gapless energy states. This provides us with another proof for the Nambu - Goldstone theorem.

5.1.9   SUMMARY OF THIS SECTION

Here is a brief summary of this section:

(1): Spontaneously broken symmetries (SBS) are generated by time independent anomalous global operators which do not annihilate vacuum.

(2): The dynamical map of these operators contains two kinds of terms; one mixes quasi-particles of the same energy and another induces a translation of gapless energy particles called the Nambu - Goldstone (NG) particles, requiring the appearance of gapless energy particles.

(3): Thus, SBS is really a change of form of the symmetry in moving from the basic level of Heisenberg fields to the observed level of quasi-particles.

(4): The translation of the NG particle fields acts as a boson transformation which regulates a NG particle condensation of the coherent state type. This makes the generator anomalous.

(5): The invariance under the NG field translation implies the low energy theorem stating that the NG fields without derivatives do not appear in dynamical maps of invariant operators.

(6): The generators are made out of the Nöther currents, for which the Ward - Takahashi relations hold.

(7): A particular form of broken symmetry is specified by an order parameter and is selected by the θ-selecting term.

5.2   Broken Phase Symmetry in a Scalar Model

5.2.1   INTRODUCTION

In this section we illustrate the general idea by a spontaneous breakdown of symmetry in a complex scalar field model. For concreteness we specify the Lagrangian, although most of the arguments and the entire strategy are applicable to any model.

5.2.2   MODEL

The Lagrangian density under consideration is

$$\mathcal{L}_N = -\partial^\mu \psi^\dagger \partial_\mu \psi + \mu^2 \psi^\dagger \psi + \lambda (\psi^\dagger \psi)^2, \tag{5.43}$$

which gives the Hamiltonian

$$H = \int d^3x [\dot{\psi}^\dagger \dot{\psi} - \nabla \psi^\dagger \nabla \psi - \mu^2 \psi^\dagger \psi - \lambda (\psi^\dagger \psi)^2]. \tag{5.44}$$

The canonical conjugate of ψ and $\psi^\dagger$ are $\dot{\psi}^\dagger$ and $\dot{\psi}$, respectively:

$$[\psi(x), \dot{\psi}^\dagger(y)]\delta(t_x - t_y) = i\delta(x - y). \tag{5.45}$$

The $\mathcal{L}_N$ is invariant under the phase transformation:

$$\psi(x) \rightarrow \psi'(x) = e^{i\theta}\psi(x), \tag{5.46}$$

which with an infinitesimal θ gives

$$\delta\psi = i\psi, \quad \delta\psi^\dagger = -i\psi^\dagger. \tag{5.47}$$

5.2.3   THE NÖTHER CURRENT AND GENERATOR

According to (5.24), the Nöther current is

$$N_\mu = -i\psi^\dagger \overset{\leftrightarrow}{\partial}_\mu \psi. \tag{5.48}$$

Here the notation $\overleftrightarrow{\partial}_\mu = \overrightarrow{\partial}_\mu - \overleftarrow{\partial}_\mu$ is used. Then, the generator is

$$N = -i \int d^3x [\psi^\dagger \overleftrightarrow{\psi} - \dot{\psi}^\dagger \psi]. \tag{5.49}$$

5.2.4 Phase Symmetry Rearrangement

The Nambu - Goldstone theorem states that there should exist a gapless energy field χ. Since we are considering a relativistic model, χ satisfies the massless equation:

$$\partial^2 \chi = 0. \tag{5.50}$$

Since the boson transformation

$$\chi \rightarrow \chi + c\theta, \tag{5.51}$$

with a c-number c, induces the phase transformation in (5.46), the dynamical map of ψ has the form [6, 8]

$$\psi = e^{i(1/c)\chi} F[\phi, \partial\chi], \tag{5.52}$$

where ϕ is a quasi-particle other than the Nambu - Goldstone particle. Since χ appears in the phase factor, it is called the phase field. Since the translation of χ takes care entirely of the phase transformation, other quasi fields, ϕ, should not change under the phase transformation. In the following we assume that we have only one quasi-particle ϕ in addition to the Nambu - Goldstone field.

However, the calculation shows that even the quasi-particle ϕ is unstable due to its decaying into the Nambu - Goldstone boson χ. Therefore it should disappear in the dynamical map when we consider higher order corrections. Thus, strictly speaking we should replace the dynamical map in (5.52) by

$$\psi = e^{i(1/c)\chi} F[\partial\chi], \tag{5.53}$$

although in a low order perturbative calculation we will find ϕ, as will be shown later.

5.2.5 The Order Parameter and θ-Selection

Let us use $(1/2)(\psi + \psi^\dagger)$ for Φ in the θ-selection term $\epsilon\Phi$, because this is not invariant under the phase transformation. Thus

$$\chi_H \equiv \delta\Phi = i\frac{1}{2}(\psi - \psi^\dagger). \tag{5.54}$$

and

$$\delta\delta\Phi = -\frac{1}{2}(\psi + \psi^\dagger). \tag{5.55}$$

Thus $\psi = \Phi - i\chi_H$; Φ is the real part of ψ, while χ_H is the imaginary part. According to (5.39), the order parameter is given by

$$v = \frac{1}{2}\langle 0|\psi + \psi^\dagger|0\rangle. \tag{5.56}$$

Thus we have the following dynamical maps:

$$\Phi = v + Z_1\phi + \cdots, \tag{5.57}$$
$$\chi_H = Z_2\chi + \cdots \tag{5.58}$$

with the c-number coefficients Z_1 and Z_2. Here $\cdots$ stands for higher order normal products. Then we can put the dynamical map (5.52) into the form

$$\psi = (v + Z_1\phi) - iZ_2\chi + \cdots, \tag{5.59}$$

We select the phase in such a manner that v is real.

5.2.6   THE DYNAMICAL MAP OF THE CURRENT

Since N generates the boson transformation (5.51) of χ, its dynamical map has the form

$$N = \int d^3x \frac{1}{c}\dot{\chi}, \tag{5.60}$$

because $\dot{\chi}$ is the canonical conjugate of χ:

$$[\chi(x), \dot{\chi}(y)]\delta(t_x - t_y) = i\delta(x - y). \tag{5.61}$$

Since N is the spatial integration of N_0, the dynamical map of N_0 is

$$N_0(x) = \frac{1}{c}\dot{\chi} + \cdots. \tag{5.62}$$

Then the Nöther current conservation law $\partial^\mu N_\mu = 0$ and the massless equation $\partial^2\chi = 0$ give the following dynamical map for the current:

$$N_\mu(x) = \frac{1}{c}\partial_\mu\chi(x) + \cdots. \tag{5.63}$$

5.2.7   A FLUCTUATION EFFECT

Since the Nambu - Goldstone field χ is of gapless energy, many low momentum χ-particles can easily be created in a short time and can sometimes cause a significant fluctuation effect. Here we study its effect on the order parameter v. According to the dynamical map (5.52) the order parameter has the form

$$v = v_s \langle 0|e^{i\frac{1}{c}\chi(x)}|0\rangle, \tag{5.64}$$

in which v_s is the contribution from $F[\phi, \partial\chi]$. Here, since we consider only the soft χ-particle contribution, we approximately replace $F[\phi, \partial\chi]$ by $F[\phi, 0]$. Thus the ϕ-effect is factorized as the v_s-factor.

Exercise 3 Derive the relation

$$\langle 0|e^{i\frac{1}{c}\chi(x)}|0\rangle = \exp\left(-\frac{1}{2c^2}D(0)\right), \tag{5.65}$$

where

$$D(0) = \langle 0|[\chi(x)]^2|0\rangle \tag{5.66}$$

$$= \frac{1}{(2\pi)^3}\int d^3k\frac{1}{2\omega_k} \tag{5.67}$$

with ω_k being the energy of χ.

Hint: Expand $\exp[i(1/c)\chi]$ in powers of χ and calculate the vacuum expectation value of each term.

When we work at a finite temperature T, then the above formula for $D(0)$ is replaced by

$$D(0) = \frac{1}{(2\pi)^3}\int d^3k\frac{1}{2\omega_k}\frac{1}{1-e^{-\beta\omega_k}} \tag{5.68}$$

with $\beta = 1/T$. (We have chosen units such that the Boltzmann constant $k_B = 1$.) Since ω_k vanishes at zero momentum, we assume $\omega_k = ak^m$ with a constant a and positive power m for small $k = |\vec{k}|$. Then, the soft χ-contribution to $D(0)$ is

$$\frac{1}{2a^2\beta(2\pi)^3}\int d^3k\frac{1}{k^{2m}}. \tag{5.69}$$

This effect is very minor in a system of three dimensions. However for a two dimensional system, d^3k is replaced by d^2k which is proportional to kdk. Then $D(0)$ contains the infrared divergence due to the small k-contribution when the power m is equal or larger than 1. For example, when $m = 1$, $D(0)$ contains the infrared divergence $\ln k$ for $k = 0$.

It is due to this divergence coming from the infrared effect that the order parameter v vanishes for the two dimensional system. The infrared divergence is enhanced to k^{-1} for $k = 0$ in a one dimensional system. Because of this infrared effect, there is no phase order of the kind studied in this section in a one or two dimensional system. This situation is not much improved even when a system has a finite size. As will be discussed in the next chapter, a system of finite size has a boundary domain which itself carries an infinite number of degrees of freedom.

It has been shown by many physicists in a variety of ways that it is difficult to create ordered states in one or two dimensional systems [12, 13, 14]. Here we introduced the arguments presented in [15, 16, 8].

However, these consideration do not exclude the appearance of short range order and order associated with discrete symmetries even in a system of lower dimension. The Kosterlitz - Thouless state [17] in two dimensional systems and solitons in one dimensional systems are some examples of this kind.

5.2.8 A STRATEGY FOR COMPUTATIONAL ANALYSIS

Calculation of c and of the mass of ϕ requires an explicit model-dependent computation. Let us begin with the crudest approximation in which the dots in the above dynamical maps are ignored. Then we have

$$\psi^\dagger\psi \;=\; (v + Z_1\phi)^2 + Z_2^2\chi^2, \tag{5.70}$$

$$\partial^\mu\psi^\dagger\partial_\mu\psi \;=\; \partial^\mu\phi\partial_\mu\phi + \partial^\mu\chi\partial_\mu\chi. \tag{5.71}$$

from which follows

$$(\psi^\dagger\psi)^2 = v^4 + 4v^3 Z_1\phi + 6v^2 Z_1^2\phi^2 + 2v^2 Z_2^2\chi^2. \tag{5.72}$$

In the last relation the terms of order higher than the second are ignored, because these terms are of the same order of magnitude as the dot terms which were ignored. Then, the linear ϕ-term in the dynamical map of the Hamiltonian density in (5.44) has the form:

$$- 2v Z_1(\mu^2 + 2v^2\lambda)\phi. \tag{5.73}$$

Since the dynamical map of the Hamiltonian must be the free Hamiltonian according to (4.26), this term should vanish, determining the order parameter v from

$$\mu^2 + 2v^2\lambda = 0, \tag{5.74}$$

because $v \neq 0$. This gives $v^2 = -(\mu^2/2\lambda)$, indicating that λ should be attractive. On the other hand, the χ^2-term is $-Z_2^2(\mu^2 + 2v^2\lambda)\chi^2$, which vanishes due to (5.74). Thus χ is massless as it should be. The ϕ^2-term is

$$- Z_1^2(\mu^2 + 6v^2\lambda)\phi^2. \tag{5.75}$$

According to (5.74) the coefficient in this term is equal to $2\mu^2 Z_1^2$, which is positive as it should be. The significance of the minus sign in front of the mass term $\mu^2\psi^\dagger\psi$ in the Hamiltonian in (5.44) in deriving these consistent answers should be noted. We see now that the dynamical map of the Hamiltonian is the sum of the free Hamiltonian of the massless χ-field and the one for the ϕ-field with the mass $2\mu^2 Z_1^2$:

$$H \;=\; \int d^3x[\dot\phi^\dagger\dot\phi + \nabla\phi^\dagger\nabla\phi + 2\mu^2 Z_1^2\phi^\dagger\phi]$$
$$+ \int d^3x[\dot\chi^2 - (\nabla\chi)^2]. \tag{5.76}$$

Thus the dynamical map of the Hamiltonian and the order parameter v are determined in this process. In a similar manner the dynamical map of the current N_μ can be obtained from (5.48) in the same approximation. The result is

$$N_\mu(x) = 2Z_2 v\partial_\mu\chi(x). \tag{5.77}$$

Comparing this with (5.63) we find $c = -(1/2Z_2v)$. We have thus calculated the ϕ-mass, the order parameter v and the coefficient c in the current N_μ in the crudest approximation. The approximation can be improved by considering loop effects in the Feynman diagrams.

As we saw above, this crude calculation finds the massive quasi-particle ϕ. However, the higher order corrections make it unstable by causing its decay to χ-particles. Thus, in a rigorous sense there is no ϕ-quasi-particle.

5.2.9 SUPERFLUID CURRENT

The dynamical map of the current shows that the superfluid current is a flow of the χ-field. Since the energy of this field is gapless, the current flow does not require any energy, leading to the superfluid phenomena. In the next chapter we will see how a condensation of χ-particle creates superfluid vortex currents.

5.3 Superconductivity

5.3.1 INTRODUCTION

Superconductivity is also a state of spontaneously broken phase symmetry. It is due to this that its dynamical mechanism shares many common features with that of superfluidity discussed in the last section. Therefore, we make this subsection rather brief. However, superconductivity differs from superfluidity in one aspect. This is the fact that superconductivity is an ordered state of charged particles such as electrons so that the electromagnetic field participates. Thus, superconductivity provides us with an example of spontaneously broken phase symmetry in a gauge theory. Since the electromagnetic field involves the Coulomb interaction among charged particles and since the Coulomb interaction has an infinite range, we may not need Nambu - Goldstone bosons in order to maintain the phase order. Indeed, gauge invariance implies that the phase becomes the longitudinal part of the electromagnetic field. Since, as we saw in the last section, the Nambu - Goldstone boson appears to be a phase field, it should become the longitudinal part of the electromagnetic field in the dynamical map of any *gauge invariant operator*. This is called the Anderson - Higgs - Kibble mechanism [18, 19, 20]. However, this does not mean that the Nambu - Goldstone boson should disappear entirely. Indeed, the general theorem, stating that the translations of χ-fields are the agents for invariant transformations associated with the broken symmetries, dictates that there should be a Nambu - Goldstone field χ even in gauge theories. The χ appears in dynamical maps of gauge noninvariant operators [15, 8], and it affects many observable results through its short time fluctuation effects which appear through internal lines in the Feynman diagrams. Superconductivity provides a simple model for this sophisticated gauge effect in phenomena of spontaneously broken symmetries.

5.3.2 THE BCS MODEL

Consider a system of an electron field $\psi_{\uparrow,\downarrow}$ with electromagnetic interaction. The suffix specifies the spin state. We choose the Coulomb gauge for the electromagnetic field so that the vector potential has only the transverse component : $\vec{\nabla}\vec{a} = 0$. Here $\vec{a}$ is the electromagnetic vector potential. To be specific we make use of the BCS-model [21], although the general strategy of the analysis can be applied to most of the models of superconductivity.

The Lagrangian density is

$$
\begin{aligned}
\mathcal{L}_N = & -\frac{1}{4}F^{\mu\nu}F_{\mu\nu} + \sum_{i=1,2}[i\psi_i^\dagger \frac{\partial}{\partial t}\psi_i \\
& -\frac{1}{2m}(\vec{\nabla} - ie\vec{a})\psi_i^\dagger(\vec{\nabla} + ie\vec{a})\psi_i + \mu_F\psi_i^\dagger\psi_i] \\
& -\lambda\psi_\uparrow^\dagger\psi_\downarrow^\dagger\psi_\downarrow\psi_\uparrow \\
& +\mathcal{L}_c,
\end{aligned}
\tag{5.78}
$$

where $\mathcal{L}_c$ is the Coulomb interaction Lagrangian, which gives the Coulomb interaction Hamiltonian

$$
H_c = \frac{1}{2}\frac{1}{4\pi}\int d^3x \int d^3y \frac{\rho(\vec{x},t)\rho(\vec{y},t)}{|\vec{x}-\vec{y}|}.
\tag{5.79}
$$

Here ρ is the charge density. Use was made also of the notation $\psi_1 = \psi_\uparrow$ and $\psi_2 = \psi_\downarrow$. The μ_F is the electron chemical potential. Note that the electromagnetic interactions appear in two ways; $\vec{a}$ appears through $(\vec{\nabla} + ie\vec{a})\psi_i$ and its Hermitian conjugate, and the Coulomb interaction operates among charges.

This Lagrangian shows that the canonical conjugate of ψ_i is $i\psi_i^\dagger$. Considering the fermionic property of electrons, we have

$$
[\psi_i(x), \psi_j^\dagger(y)]_+\delta(t_x - t_y) = \delta(x - y),
\tag{5.80}
$$

where use was made of the notation $[A, B]_+ = AB + BA$.

The Lagrangian is invariant under the phase change $\psi_i \rightarrow \exp[i\theta]\psi_i$, which with an infinitesimal θ gives

$$
\delta\psi_i = i\psi_i, \quad \delta\psi_i^\dagger = -i\psi_i^\dagger.
\tag{5.81}
$$

The Lagrangian is invariant also under the gauge transformation:

$$
\vec{a}(x) \rightarrow \vec{a}(x) + \vec{\nabla}\lambda(x), \quad \psi_i(x) \rightarrow e^{ie\lambda(x)}\psi_i(x)
\tag{5.82}
$$

Here, *the gauge function is required to have its Fourier transform exist.*

5.3.3 THE CURRENT AND THE GENERATOR

The electric current $\vec{j}$ and charge density ρ are defined through the Maxwell equation. They are given by the Nöther current N_μ for the electron phase transformation as follows:

$$\vec{j} = -e\vec{N}, \quad \rho = -eN_0. \tag{5.83}$$

According to (5.24) and (5.81), the current, the charge density and the generator are

$$\vec{j} = \sum_{i=1,2} [-\frac{ie}{2m}\psi_i^\dagger \overset{\leftrightarrow}{\nabla} \psi_i + \frac{e}{m}\vec{a}\psi_i^\dagger \psi_i], \tag{5.84}$$

$$\rho = \sum_{i=1,2} \psi_i^\dagger \psi_i, \tag{5.85}$$

$$N = \frac{1}{e} \int d^3x \rho(x). \tag{5.86}$$

Here the notation $\overset{\leftrightarrow}{\nabla} = \overset{\rightarrow}{\nabla} - \overset{\leftarrow}{\nabla}$ is used.

5.3.4 PHASE SYMMETRY REARRANGEMENT

The Nambu - Goldstone theorem states that there should exist a gapless energy field χ. Since the boson transformation

$$\chi \to \chi + c\theta \tag{5.87}$$

with a c-number c induces the phase transformation in (5.46), the dynamical map of $\psi_{\uparrow,\downarrow}$ has the form [6, 8]

$$\psi_{\uparrow,\downarrow} = e^{i\frac{1}{\eta(\nabla^2)}\chi} F_{\uparrow,\downarrow}[\varphi, \vec{a} - \frac{1}{e}\eta^{-1}(\nabla^2)\vec{\nabla}\chi], \tag{5.88}$$

where φ is the quasi electron and $\eta(\nabla^2)$ is a derivative operator such that $\eta(0) = (1/c)$. Since χ appears in the phase factor, it is called the phase field. As $\eta^{-1}(\nabla^2)\chi$ is the electron phase, the gauge transformation is given by $\vec{a} \to \vec{a} + \vec{\nabla}\lambda$ and $\chi \to \chi + e\eta(\nabla^2)\lambda$ with a Fourier transformable c-number $\lambda(x)$. Thus gauge invariance dictates that $\vec{a}$ and $\vec{\nabla}\chi$ in the dynamical map of any gauge invariant operator should appear through the combination $[\vec{a} - \frac{1}{e}\eta^{-1}(\nabla^2)\chi]$. This was taken into account by the above form for the dynamical map of $\psi_{\uparrow,\downarrow}$. Since translation of χ takes full care of the phase transformation, the quasi electron field φ should not change under the phase transformation. This is an example of freezing phenomena.

5.3.5   THE DYNAMICAL MAP OF CHARGE DENSITY

Since N generates the boson transformation (5.87) of χ, its dynamical map
has the form

$$N = \int d^3x \eta(\nabla^2)\pi(x), \tag{5.89}$$

where π is the canonical conjugate of χ:

$$[\chi(x), \pi(y)]\delta(t_x - t_y) = i\delta(x - y). \tag{5.90}$$

Since N is the spatial integration of N_0, the dynamical map of charge
density is

$$\rho(x) = -e\eta(\nabla^2)\pi(x) + \cdots, \tag{5.91}$$

where dots stand for higher order normal products.

5.3.6   THE NAMBU - GOLDESTONE FIELD

The free field equation for χ is not as simple as the one in the case of
superfluidity, because π is not $\dot\chi$. This complication is caused by the fact
that the Coulomb interaction contains π.

When the Coulomb interaction is ignored, the Hamiltonian of χ has the
form

$$H_\chi^0 = \frac{1}{2}\int d^3x \left(\pi^2 + [v_B(\nabla^2)\vec\nabla\chi]^2\right). \tag{5.92}$$

Here, $v_B(\nabla^2)$ is defined by the boson energy ω_k as $\omega_k^2 = v_B(-k^2)k^2$. When
we add the Coulomb interaction to this and consider (5.91), we obtain the
Hamiltonian

$$\begin{aligned}
H_\chi &= \frac{1}{2}\int d^3x \left(\pi^2 + [v_B(\nabla^2)\vec\nabla\chi]^2\right) \\
&+ \frac{1}{2}\frac{e^2}{4\pi}\int d^3x \int d^3y \eta(\nabla^2)\pi(\vec x,t)\frac{1}{|\vec x - \vec y|}\eta(\nabla^2)\pi(\vec y,t).
\end{aligned} \tag{5.93}$$

The commutation relation between this Hamiltonian and χ or π gives
the following field equations:

$$\begin{aligned}
\dot\chi(x) &= \pi(x) + \frac{e^2}{4\pi}\eta(\nabla^2)\int d^3y \frac{1}{|\vec x - \vec y|}\eta(\nabla^2)\pi(\vec y,t), & (5.94)\\
\dot\pi(x) &= v_B^2(\nabla^2)\nabla^2\chi(x). & (5.95)
\end{aligned}$$

This together with the current conservation law

$$\frac{\partial}{\partial t}\rho + \vec\nabla\vec j = 0 \tag{5.96}$$

(which follows from the conservation law of the Nöther current) and the structure of charge density in (5.91) leads to the following linear χ-term in the dynamical map for the current:

$$\vec{j}_\chi = -e\eta(\nabla^2)v_B^2(\nabla^2)\vec{\nabla}\chi. \tag{5.97}$$

According to (5.88) $\vec{a}$ and χ appear in the dynamical map of any gauge invariant operator through the combination $[e\vec{a} - \eta^{-1}(\nabla^2)\vec{\nabla}\chi]$. Since the current is gauge invariant, we find the following dynamical map for the current:

$$\vec{j} = e^2\eta^2(\nabla^2)v_B^2(\nabla^2)[\vec{a} - \frac{1}{e}\eta^{-1}(\nabla^2)\vec{\nabla}\chi] + \cdots. \tag{5.98}$$

The fact that $\eta(0)v_B(0)$ does not vanish is the origin of the Meissner current. This will be discussed in the next section.

The presence of this linear $\vec{a}$-term in the current means that the free Hamiltonian of the quasi photon has the term

$$\frac{1}{2}e^2\vec{a} \cdot \eta^2(\nabla^2)v_B^2(\nabla^2)\vec{a}, \tag{5.99}$$

because $\vec{j} = -(\partial H/e\partial\vec{a})$. This implies that a photon in a superconducting system has energy $\omega_{pl}(k) = e\eta(-k^2)v_B(-k^2)$, which does not vanish at zero momentum. This is the transverse plasma in superconducting states.

In relativistic cases this means that the gauge field acquires mass due to spontaneously broken symmetry. This is one aspect of the Anderson - Higgs - Kibble mechanism [18, 19, 20]. Another aspect will be discussed in the next subsection.

5.3.7 ARE NG-BOSONS ENTIRELY ELIMINATED?

The two equations (5.94) and (5.95) lead to

$$\left(\frac{\partial^2}{\partial t^2} - v_B^2(\nabla^2)\nabla^2\right)\chi =$$
$$\frac{e^2}{4\pi}\eta(\nabla^2)\int d^3y\frac{1}{|\vec{x} - \vec{y}|}\eta(\nabla^2)v_B^2(\nabla^2)\nabla^2\chi(\vec{y},t). \tag{5.100}$$

This is the equation for the NG boson field. Note that *we cannot apply integration by parts to the last term because $1/|\vec{x} - \vec{y}|$ has infinite range.* One careful approach to this integration is to put it in the Fourier form and to introduce the low momentum cut-off k_c in the integration with the Coulomb potential. With this cut-off we can safely make use of integration by parts and obtain

$$\left[\frac{\partial^2}{\partial t^2} - v_B^2(\nabla^2)\nabla^2 + e^2\eta^2(\nabla^2)v_B^2(\nabla^2)\right]\chi = 0 \tag{5.101}$$

for χ with momentum $|\vec{k}| > k_c$. Since there is no Coulomb term for χ with lower momentum, we have

$$\left[\frac{\partial^2}{\partial t^2} - v_B^2(\nabla^2)\nabla^2\right]\chi = 0 \qquad (5.102)$$

for $|\vec{k}| < k_c$. At the end of calculations the limit $k_c \to 0$ should be performed. (Roughly speaking, k_c is of order of magnitude about equal to the inverse of linear size of the system.) In this limit, χ becomes independent of time and the equation (5.102) is replaced by the Laplace equation

$$\nabla^2\chi = 0, \qquad (5.103)$$

because this picks up the zero momentum contribution of χ.

Thus, at the limit of vanishing k_c we have two equations; (5.101) and (5.103). The solution of the equation (5.101) with low momentum has the energy $\omega_{pl} = e\eta(0)v_B(0)$. This is the plasma energy. The χ-part satisfying (5.101) gives the longitudinal plasmon.

As mentioned above, the Fourier transformable part of the solution of the Laplace equation (5.103) picks up only the zero momentum contribution. We may state that the Coulomb interaction brings up the Nambu - Goldstone mode to the plasma mode for the entire momentum domain *except the zero momentum* [15, 8]. Therefore, only the observable boson excitation levels are the plasma levels; the gapless excitation levels (the phason levels) are eliminated from observation by the Coulomb effect. This is called the Anderson - Higgs - Kibble mechanism [18, 19, 20].

However, the phase field satisfying the Laplace equation (5.103) cannot be ignored, because *its presence is required by the phase invariance* [15, 8]. The phase transformation of the electron field is generated by the translation of the χ, which requires the zero momentum and zero energy part of the NG field. Though this boson is not observed as an excitation level, this is the agent for the phase transformation.

Furthermore, we will see in the next chapter that *this solution of the Laplace equation is the origin of the vortices* which appear in mixed states of superconductors.

The fluctuation effects due to this soft χ sometimes have an important effect. We will illustrate this in the following subsections by studying the order parameter.

5.3.8   ORDER PARAMETER AND θ-SELECTION

Let us use the following choice for Φ in the θ-selection term $\epsilon\Phi$:

$$\Phi(x) = \frac{1}{2}[\psi_\uparrow(x)\psi_\downarrow(x) + \psi_\downarrow^\dagger(x)\psi_\uparrow^\dagger(x)]. \qquad (5.104)$$

This gives

$$\chi_H \equiv \delta\Phi \tag{5.105}$$

$$= i\frac{1}{2}[\psi_\uparrow(x)\psi_\downarrow(x) - \psi_\downarrow^\dagger(x)\psi_\uparrow^\dagger(x)]. \tag{5.106}$$

and

$$\delta\delta\Phi = -\Phi \tag{5.107}$$

Thus Φ is the real part of $\psi_\uparrow(x)\psi_\downarrow(x)$, while $-\chi_H$ is the imaginary part. According to (5.39), the order parameter is given by

$$v = \frac{1}{2}\langle 0|\Phi|0\rangle. \tag{5.108}$$

Thus we have the following dynamical map:

$$\chi_H = Z(\nabla^2)\chi + \cdots. \tag{5.109}$$

Here the Z may be a function of the spatial derivative, because the model is not relativistic. We select the phase in such a manner that v is real.

The energy spectrum of the NG boson and the derivative operator $Z(\nabla^2)$ are obtained from calculation of the function $\langle 0|T[\chi_H(x)\chi_H(y)]|0\rangle$.

5.3.9   A FLUCTUATION EFFECT

Following the computation in subsection 5.2.7, we obtain the following result for the order parameter:

$$v = v_s \exp\left(-\frac{1}{c^2}D(0)\right), \tag{5.110}$$

with

$$D(0) = \frac{1}{(2\pi)^3}\int d^3k\,\theta(k_c - |\vec{k}|)\frac{1}{2\omega_k}\frac{1}{1 - e^{-\beta\omega_k}} \tag{5.111}$$

and $\beta = 1/T$. Here we ignored the plasma effect because it does not alter the infrared effect. Therefore, the momentum integration is confined to the domain $|\vec{k}| \leq k_c$. The energy of χ with the momentum less than k_c is $\omega_k = v_B(k^2)k$. Thus, this result is the same as the superfluid case discussed in subsection 5.2.7 except that the momentum integration is confined to the momentum less than k_c. Hence, its effect in a system of three dimensions is minor. However, the instability of the phase order in two or one dimensional systems discussed in subsection 5.2.7 is applicable to the superconducting phase order when k_c is finite, however small k_c is. Therefore this instability remains true even in the limit of vanishing k_c. Thus, we state that the phase ordered state of the kind considered here is unstable in two or one dimensional systems [15, 16, 8].

There are materials which contain some kind of molecular arrangements forming two or one dimensional structures. A famous set are the copper - oxide high critical temperature superconductors. Since these materials contain many two dimensional sheets, tunneling effects between sheets and lines may avoid the instability considered here. This is just one possibility, there could be other mechanisms which could avoid this instability. Also if this tunneling mechanism is working we would expect strong fluctuation effects in these materials.

Although the general mechanism of spontaneous breakdown of phase symmetry is simple as has been seen in this section, actual manifestations can exhibit a rich variety of behavior, depending on the structure of the samples.

5.3.10   GAUGE INVARIANCE

We stated that the gauge invariance of the theory dictates the gauge rule that the dynamical map of any gauge invariant operator should contain χ through the combination $[e\vec{a}-\eta^{-1}(\nabla^2)\vec{\nabla}\chi]$. This requires a short comment. Since we specified the vector potential $\vec{a}$ to be transverse in order to use the Coulomb gauge and since $\vec{\nabla}\lambda(x)$ with a Fourier transformable λ is longitudinal, we have no freedom for gauge transformations of $\vec{a}$.

Indeed, since the canonical conjugate of $\vec{a}$ is its time derivative, the generator for the gauge transformation of $\vec{a}$ is

$$N_a = \int d^3x \frac{\partial}{\partial t}\vec{a}(x) \cdot \vec{\nabla}\lambda(x), \qquad (5.112)$$

which vanishes because $\vec{\nabla}\vec{a} = 0$.

However we can use the Coulomb gauge without specifying the vector potential to be transverse. Indeed, the general form of vector potential $\vec{A}$ in the Coulomb gauge is

$$\vec{A}(x) = \vec{a}(x) + \vec{\nabla}\lambda(x) \qquad (5.113)$$

with a c-number function $\lambda(x)$. Here $\vec{a}$ is the transverse vector potential operator.

Now recall the c-q transmutation rule presented in section 1.4 in chapter 1. This rule dictates that any c-number change should be created by a q-number change. Hence, any change of the c-number $\lambda(x)$ should be generated by a change of the q-number.

On the other hand, since the gauge transformation changes the electron phase by $e\lambda(x)$, it should make the change $\chi(x) \rightarrow \chi(x) + e\eta(\nabla^2)\lambda(x)$, because $\eta^{-1}(\nabla^2)\chi$ is the electron phase. In other words gauge invariance demands that this change of χ should correspond to the change of the c-number $\lambda(x)$ in the vector potential $\vec{A}(x)$. This leads to the above gauge rule.

5.3.11 A STRATEGY IN COMPUTATIONAL ANALYSIS

Analysis of superconducting materials requires calculation of electric and magnetic quantities. The magnetic properties are mostly controlled by the structure of the Meissner current which was obtained in (5.98).

It is common to rewrite the form (5.98) for the current as

$$\vec{j} = -\frac{1}{4\pi\lambda_L^2}c(\nabla^2)[\vec{a} - \frac{1}{e}\eta^{-1}(\nabla^2)\vec{\nabla}\chi] + \cdots \qquad (5.114)$$

with the condition

$$c(0) = 1. \qquad (5.115)$$

This determines λ_L to be

$$\frac{1}{\lambda_L^2} = 4\pi e^2 v_B(0)\eta^2(0). \qquad (5.116)$$

The constant λ_L is called the *London penetration length*. The $c(\nabla^2)$ in (5.114) is the form factor of the Meissner current. The range of this function is called the *coherence length* and is usually denoted by ξ. The ratio $\kappa = 0.96\lambda_L/\xi$ is called the Ginzburg - Landau parameter. To simplify notations it is common to use $\kappa_B = \lambda_L/\xi$ which is approximately the same as κ. Many significant quantities are controlled by these constants including the energy spectrum of quasi electrons.

The superconducting state is related to the normal conducting state through a fermionic Bogoliubov transformation, which induces a condensation of spin up and down electron pairs called Cooper pairs. In this way the superconductor Fock space is inequivalent to the normal conductor Fock space. The calculation of energy of quasi electron is made by help of this Bogoliubov transformation which creates condensation of the Cooper pairs.

The calculations of λ_L, ξ and $c(\nabla^2)$ require the computation of the two point functions $\langle 0|T[\chi_H(x)\chi_H(y)]|0\rangle$, because these quantities are controlled by the dynamical behavior of the NG bosons. These quantities have been calculated in the random phase approximation for the BCS-model [18, 15, 8].

The persistent current is the current generated by the condensation of χ satisfying the Laplace equation. Since this field does not carry any energy, the current can flow without resistivity, leading to the superconducting current. The two point propagator of the χ_H field carries the zero energy singularity associated with the NG boson χ, leading to the infinite conductivity. (See, for example, page 352 in [8].)

Although each material requires a particular method of analysis depending on its characteristic nature, the general consideration such as the one in this section frequently provides a useful guideline in practical analysis.

5.4 Non-Abelian Symmetry Breakdown

5.4.1 SPONTANEOUSLY BROKEN SPIN SYMMETRY

In the last two sections we studied a spontaneously broken phase symmetry which is an Abelian symmetry. In this section we consider a macroscopic spin state which is created by spontaneous breakdown of the spin rotational symmetry which is non-Abelian.

Let S_i with $i = 1, 2, 3$ denote the three operators forming the SU(2) algebra:

$$[S_i, S_j] = i\epsilon_{ijk}S_k, \tag{5.117}$$

where ϵ_{ijk} is the totally antisymmetric tensor with the property $\epsilon_{123} = 1$. The generators are then given by $S_i - \langle 0|S_i|0 \rangle$. We assume that the vacuum is invariant under rotation around the third axis. Since S_1 and S_2 form a two dimensional vector in the plane orthogonal to the third direction, we have $\langle 0|S_i|0 \rangle = 0$ for $i = 1, 2$.

The local spin operators will be denoted by $s_i(x)$:

$$S_i = \int d^3x \; s_i(x), \quad i = 1, 2, 3. \tag{5.118}$$

Then, the operators s_i behave as a triplet vector under the spin rotation.

We choose s_3 for Φ in the θ-selecting term $\epsilon\Phi$. Since s_3 is the third component of the triplet vector, we have

$$\delta_1\delta_1 s_3 = \delta_2\delta_2 s_3 = -s_3, \tag{5.119}$$

where δ_i means the change generated by S_i. Thus, according to (5.39) the order parameter is

$$v = \langle 0|s_3|0 \rangle, \tag{5.120}$$

which is the macroscopic spin momentum density pointing the third direction.

The generator for the rotation around the third axis is $S_3 - \langle 0|S_3|0 \rangle$. Since this symmetry is not broken in the vacuum $|0 \rangle$, we have

$$[S_3 - vV]|0 \rangle = 0. \tag{5.121}$$

Here V is given by $\langle 0|S_3|0 \rangle = vV$, implying vV is the total spin and therefore, V is the volume of the system. Note that, since the vacuum expectation of S_3 does not vanish, S_3 does not annihilate the vacuum $|0 \rangle$. Thus, as was shown in section 4.5, the state $S_3|0 \rangle$ does not have a finite norm; rather the state $[S_3 - vV]|0 \rangle$ is the one whose norm is finite. Thus, appearance of V in the above equation is not surprising.

Invariances other than the third one are broken:

$$S_i|0 \rangle \neq 0, \quad i = 1, 2. \tag{5.122}$$

Thus we have a case of spontaneous breakdown of SU(2) symmetry with a U(1) subgroup symmetry left unbroken. The subgroup associated with the unbroken symmetry is called the stability group. This provides an example of spontaneous break down of SU(2) symmetry with U(1) stability group.

Since the S_i-symmetries with $i = 1, 2$ are broken in the vacuum, these operators generate translation of the NG boson χ. To analyze this translation it is convenient to introduce

$$S_\pm \equiv S_1 \pm iS_2. \tag{5.123}$$

These operators induce translation of the χ-field according to the general consideration of the symmetry rearrangement in subsection 5.1.3. We write the translation induced by S_- as

$$\chi(x) \rightarrow \chi(x) + c\theta \tag{5.124}$$

with the transformation parameter θ. The phase of the complex c-number c depends on the direction of the (1-2)-axes in the (1-2)-plane. Therefore, without loss of generality we can choose c to be real.

In ferromagnetism χ usually has the form

$$\chi(x) = \frac{1}{(2\pi)^{3/2}} \int d^3k \chi_k e^{i[\vec{k}\cdot\vec{x}-\omega_k t]}. \tag{5.125}$$

Then we have

$$[\chi(x), \chi^\dagger(y)]\delta(t_x - t_y) = \delta(x - y), \tag{5.126}$$

which implies that $i\chi^\dagger$ is the canonical conjugate of χ. Then, the above χ-translation generated by S_- indicates that the dynamical map of S_- is

$$S_- = c \int d^3x \chi^\dagger(x). \tag{5.127}$$

The Hermitian conjugate of this gives the dynamical map of S_+. The NG field χ is called the magnon or spin wave.

The relation (5.117) gives

$$[S_3, S_\pm] = \pm S_\pm. \tag{5.128}$$

This shows that S_3 changes $S_\pm$ only by a phase, implying the dynamical map

$$S_3 = vV + \int d^3k \chi_k^\dagger \chi_k. \tag{5.129}$$

Note that the above dynamical maps give

$$[S_+, S_-] \overset{w}{=} c^2 V. \tag{5.130}$$

This is a weak relation, meaning that the equation (5.130) holds true only when we consider its matrix elements among state vectors in the Fock space built on the vacuum $|0\rangle$.

On the other hand, the vacuum expectation values of both sides of the equation (5.117) gives a relation which also contains V explicitly:

$$\langle 0|[S_+, S_-]|0\rangle = 2vV. \tag{5.131}$$

This determines the constant c in terms of the spin density v as

$$c^2 = 2v \tag{5.132}$$

The reason for the change between the weak relation (5.130) and the relations in (5.117) is the following [22]. The vacuum expectation value of S_i is given by spatial integration of $\langle a|s_i(x)|b\rangle$ with $|a\rangle$ and $|b\rangle$ standing for any two state vectors belonging to the Fock space built on $|0\rangle$. The point here is that the matrix elements are calculated before the spatial integration. When these matrix elements contain terms of order $O(1/V)$ with the volume V this should be regarded as zero at the limit of infinite V. Then these terms are missed even after the integration, leading to the weak relation (5.130). These terms would be recovered if the integration were performed before the limiting process, leading to the relation (5.117).

The case of a spontaneously broken SU(2)-symmetry is much simpler in a relativistic case. Consider a Lagrangian for a triplet isovector ψ_i with $i = 1, 2, 3$. Each component is a Hermitian Heisenberg operator. Assume that the Lagrangian is invariant under the isospin rotation; thus it has the SU(2)-symmetry.

We assume that this rotational symmetry is broken in the vacuum, although the rotational symmetry around the third axes is preserved:

$$S_i|0\rangle \neq 0, \quad i = 1, 2. \tag{5.133}$$
$$S_3|0\rangle = 0. \tag{5.134}$$

Since there are two generators associated with broken symmetries, there exist two NG fields $\chi_i : i = 1, 2$ such that *the translation*

$$\chi_i \rightarrow \chi_i + c\theta \tag{5.135}$$

induces the transformation $\exp[-i\theta S_i]$. The c-number constant c does not depend on i because S_3-symmetry is not broken.

Since the two translations in (5.135) commute with each other,

$$[S_1, S_2] \overset{w}{=} 0. \tag{5.136}$$

This is again a weak relation. The reason for the difference between this relation and the relations in (5.117) is the same as the one in the ferromagnetic model discussed above.

Since the rotation symmetry around the third axis is not broken, the operators S_i behave as two components of a two dimensional vector:

$$[S_3, S_i] = i\epsilon_{3ij}S_j \tag{5.137}$$

for $i = 1, 2$.

Summarizing, the observed symmetry consists of two χ-translations and one rotation forming the two dimensional Euclidean E(2) algebra [22].

In relativistic models the canonical conjugate of χ_i is its time derivative $\dot{\chi}_i$:

$$[\chi_i(x), \dot{\chi}_j(y)]\delta(t_x - t_y) = i\delta_{ij}\delta(x - y). \tag{5.138}$$

Thus, the dynamical maps are

$$S_i = \frac{1}{c}\int d^3x \dot{\chi}_i(x), \quad i = 1, 2 \tag{5.139}$$

$$s_i(x) = \frac{1}{c}\dot{\chi}_i(x) + \cdots, \quad i = 1, 2 \tag{5.140}$$

$$S_3 = \int d^3x[\dot{\chi}_1\chi_2 - \dot{\chi}_2\chi_1]. \tag{5.141}$$

In the dynamical map of S_3 we wrote only the contributions of the χ-fields. There can be other quasi-particles which could form irreducible representations of the E(2) group mentioned above. The NG fields χ_i are massless.

5.4.2 SYMMETRY REARRANGEMENT

In sections 5.2 and 5.3 we saw that the phase symmetry, which is Abelian, is rearranged to the NG-boson translation, which is also Abelian. However, the spin model in the last subsection showed us a surprising fact that even the group property can change through symmetry rearrangement.

The relativistic isospin model gave an example of SU(2)-basic symmetry rearranged to E(2) observed symmetry. This illustrates two features of symmetry rearrangement. One is that the number of generators does not change through the rearrangement. This is obvious but is frequently overlooked. For example, in case of ferromagnetism, only the rotational symmetry around the third axis is easily visible and this could give people the wrong idea that the symmetry has been reduced. As we saw in the last section, the symmetry is not reduced because both the basic and observable symmetries have the same number of transformations. Another feature of symmetry rearrangement is that every broken symmetry induces a translation of the NG boson fields. This second point frequently leads to a change in group structure, because the field translations are Abelian in nature.

The reason for the group change is due to disappearance of the $(1/V)$-term in matrix elements of the generator density. The limit $V \to \infty$ is to be performed before the spatial integration [22], and since the group change occurs at the limit of the volume parameter, this is a kind of group deformation.

Those readers, who are familiar with group contraction [23], may have noticed that the change SU(2) $\to$ E(2) is a group contraction.

The symmetry rearrangement in a real scalar field model of the symmetry group SU(2) × SU(2) with the stability group (maximum unbroken-symmetry subgroup) takes the form SU(2)×SU(2) → E(3), where E(3) is the three dimensional Euclidean algebra. This is again a group contraction. An example of this kind is the SU(2)×SU(2) chiral symmetry for pion physics. This symmetry is spontaneously broken with the pion being the triplet NG boson, following the rearrangement mentioned above: SU(2)×SU(2) → E(3). This is quite significant. For example, if the observed symmetry were SU(2)×SU(2), it would not be able to accommodate a triplet pion, but E(3) can do it.

It has been shown that the complete condition for a symmetry rearrangement to be a group contraction is that the stability group be the maximal subgroup. (See page 307 in [8])

Although not all of the symmetry rearrangements are group contractions, they can be formulated by the group deformation, as it was shown by Weimar and Joos [24, 25].

5.4.3   LOW ENERGY THEOREM II

The simplest form of the low energy theorem associated with the NG boson was presented in section 5.1.5. The theorem stated that the soft NG-bosons do not participate in any reaction. In this subsection, we present a more complicated low energy theorem. This will illustrate how a knowledge of the structure of symmetry rearrangement helps us to find certain relations among reactions in which some soft NG-bosons participate.

Consider a Heisenberg operator $F[\psi]$ which is a sum of products of Heisenberg fields ψ. Assume that some symmetries are spontaneously broken and give rise to the NG-boson fields $\chi_i; i = 1, 2, \cdots$. The dynamical map of F is denoted by $F[\chi_1, \chi_2, \cdots]$, in which we do not explicitly write quasi-particle fields other than the NG-bosons.

Consider the generator N_i whose action is generated by translation of the NG-boson χ_i:

$$\chi_i \rightarrow \chi_i + c_i \theta \tag{5.142}$$

with the transformation parameter θ. The change of F due to infinitesimal N_i-transformation divided by the infinitesimal parameter θ is denoted by $\delta_i F$. Let us focus our attention on the N_1-transformation:

$$\delta_1 F[\psi] = \frac{1}{\theta} \left(F[\chi_1 + c_1\theta, \chi_2, \cdots] - F[\chi_1, \cdots] \right). \tag{5.143}$$

Exercise 4 Derive from the above relation the formula

$$\langle a|\delta_1 F[\psi]|b\rangle = \lim_{k\to 0} (2\pi)^{3/2} [2\omega_k^1]^{1/2} c_1 \langle a|F[\psi]|b, \chi_1(k)\rangle \tag{5.144}$$

$$= \lim_{k\to 0} (2\pi)^{3/2} [2\omega_k^1]^{1/2} c_1 \langle a, \chi_1(k)|F[\psi]|b\rangle. \tag{5.145}$$

Here assume the following form for the NG-boson fields:

$$\chi_i(x) = \int d^3k \frac{1}{[(2\pi)^3 2\omega_i(k)]^{(1/2)}} \left(\chi_{i,k} e^{i[\vec{k}\cdot\vec{x}-\omega_i(k)t]} + \chi_{i,k}^\dagger e^{-i[\vec{k}\cdot\vec{x}-\omega_i(k)t]} \right).$$

$$(5.146)$$

The symbol $|b, \chi_i(k)\rangle$ means the state consisting of $|b\rangle$ and the χ_i-boson with the momentum $\vec{k}$.

In [26] a detailed account of this low energy theorem was presented. There, the theorem was applied to several reactions. Its application to the chiral SU(2)×SU(2) theory for pion physics provided a new derivation of the so called Callan - Treiman relations for the K-lepton decay; the relations between K→ $l\pi$ and K→ $l2\pi$ decays. Here l means a lepton. It was shown that many results of current algebra theory can be obtained from symmetry rearrangement.

5.5 c-q Transmutation and Crystal Phonon

Up to now we have considered only those ordered states which are homogeneous in space. However, in reality there is a huge variety of ordered states and many of them have some spatial structure. When a Lagrangian is given, it is hard to guess what kinds of ordered states are permitted. Usually analysis begins with some kinds of intuitive or physical reasons suggesting certain possible choices to us. Assuming that such an ordered state is realizable, we try to see if it is consistent with the basic equations obtained from the Lagrangian. This is how many analyses proceed.

In this section, as a simple example of a spatially inhomogeneous order, we consider very briefly a cubic crystal order. Let $\psi(x)$ denote a molecular field. Since the molecule number is conserved, the Lagrangian should be invariant under the phase transformation $\psi \to \exp[i\theta]\psi$. Suppose that, as happens in many nonrelativistic models, $\psi^\dagger$ is the canonical conjugate of ψ. Then the generator for the phase transformation is

$$N = \int d^3x \rho(x) \qquad (5.147)$$

with

$$\rho(x) = \psi^\dagger(x)\psi(x). \qquad (5.148)$$

This ρ is the fourth component of the Nöther current associated with the phase transformation, and therefore, is the molecule number density.

However, the most significant invariance under consideration is not the phase invariance but the spatial translation invariance. The Lagrangian is invariant under the spatial translation $\vec{x} \to \vec{x} + \vec{\theta}$ with the continuous parameter θ. The total momentum operator $\vec{P}$ is the generator for this transformation.

Now, define

$$n(\vec{x}) = \langle 0|\rho(x)|0\rangle. \tag{5.149}$$

This is the macroscopic molecule number density. The order parameter is then given by

$$\vec{v}(\vec{x}) \equiv \vec{\nabla} n(\vec{x}) \neq 0. \tag{5.150}$$

This condition shows that the order parameter does depend on $\vec{x}$ so that it is not homogeneous.

In the case of a cubic crystal we have a further condition stating that $n(\vec{x})$ is a periodic function with three orthogonal lattice vectors $\vec{a}_i$ ($i = 1, 2, 3$):

$$n(\vec{x} + \vec{a}_i) = n(\vec{x}). \tag{5.151}$$

A particularly interesting aspect of this subject is that the space translation is a transformation, not of an operator, but of the c-number position $\vec{x}$. As was explained in section 1.4, the c-q transmutation theorem states that this translation of $\vec{x}$ should be generated by a transformation of a certain operator.

It is easy to identify this operator. Since the spatial translation symmetry is now spontaneously broken, there should exist gapless energy NG bosons. Since there are three degrees of freedom of translation, there appear three NG boson fields χ_i; $i = 1, 2, 3$, which are mutually commutable and which form a vector $\vec{\chi}$. The spatial translation should be taken care of by translation of these NG-boson fields. However, since $\vec{\chi}$ itself depends on $\vec{x}$, this position should also be translated. Thus, the NG-boson translation, which generates the spatial translation, is

$$\vec{\chi}(\vec{x}, t) \rightarrow \vec{\chi}(\vec{x} + \vec{\theta}) + c\vec{\theta} \tag{5.152}$$

with a constant c.

The dynamical map of $\vec{P}$ which generates this transformation is

$$P_i = \int d^3k k_i (\vec{\chi}_k^\dagger \cdot \vec{\chi}_k) + \frac{1}{c} \int d^3x \dot{\chi}_i(x), \tag{5.153}$$

where we have considered the cases in which $\dot{\chi}_i$ is the canonical conjugate of χ_i as it is in the case of most crystals. The first term in P_i generates the spatial translation of $\vec{x}$ in $\chi_i(x)$, while the second term should generate the translation of χ_i. Now the c-q transmutation theorem tells us that the latter, the translation of NG-bosons, should induce the translation of $\vec{x}$ in $n(\vec{x})$. This means that both $\vec{x}$ and $\vec{\chi}$ should appear in pertinent expressions. For example, the dynamical map of the molecule number density has the form

$$\begin{aligned} \rho(x) &= F[\vec{x} + (1/c)\vec{\chi}(x), \partial \vec{\chi}] \tag{5.154} \\ &= n(\vec{x}) + \frac{1}{c}\vec{\chi}(x) \cdot \vec{\nabla} n(\vec{x}) + \cdots. \tag{5.155} \end{aligned}$$

A systematic quantum field formalism for the analysis of crystals has been presented in [27, 28, 29, 8]. A systematic application of this formalism to point defect energy in a series of crystals and comparison of its results to experimental data were presented in [30].

The above NG-bosons are the acoustic phonons. The derivation of these phonons on basis of the c-q transmutation rule is a useful example which illustrates quite a general rule for spontaneous breakdown of spatial translation symmetry. In the next chapter creation of macroscopic objects due to boson condensation will be studied. There again, spatial translation symmetry is broken by the presence of macroscopic objects and the c-q transmutation rule gives rise to zero energy modes. However, contrary to the situation in this section, these zero energy modes are not particle modes but discrete zero energy modes. This is because these zero energy modes have nothing to do with lattice oscillations, but reflect the fact that the choice of the position of the center of mass of extended objects is arbitrary; a shift in the center of mass point does not consume any energy, leading to the zero energy mode.

5.6   The Spontaneous Creation of Mass

Consider a model which is invariant under the scale transformation:

$$\psi(x) \rightarrow \psi'(x) = e^{\theta}\psi(e^{\theta}x) \tag{5.156}$$

with a real parameter θ.

A simple model of this kind is given by the Heisenberg equation

$$\partial^{\mu}\partial_{\mu}\psi(x) = \lambda\psi^3(x), \tag{5.157}$$

where ψ is a Hermitian field and λ is a dimensionless coupling constant.

Note that ψ acquires the same scaling factor $\exp\theta$ as the one for x, although its dimension is inverse of length. This can be understood as follows:

$$\partial^{\mu}\partial_{\mu}\psi'(x) = e^{3\theta}\partial_z^{\mu}\partial_{z\mu}\psi(z) \tag{5.158}$$
$$= e^{3\theta}\lambda\psi^3(z) \tag{5.159}$$
$$= \lambda\psi'^3(e^{\theta}x), \tag{5.160}$$

where $z = e^{\theta}x$

The dimension of ψ is determined by the canonical commutation relation which shows that the commutation relation between $\psi(x)$ and $\dot{\psi}(y)$ with $t_x = t_y$ is proportional to $\delta(\vec{x} - \vec{y})$, whose dimension is the inverse of the third power of the length. This implies that the dimension of ψ is the inverse of length.

Note that this model has no parameters with dimension of length. In the early days of quantum field theory it was said that there should be no massive quasi-particle in this case, simply because there is no dimensional parameter out of which a mass can be created. However, the discovery of the mechanism for spontaneously broken symmetries changed the situation. With a choice of vacuum which is not scale invariant we can have massive quasi-particles. The main question to be asked is how the c-q transformation condition is satisfied; what kind of q-number transformation induces the mass scaling $m \rightarrow \exp[-\theta]m$? We briefly touch on this question in this section.

The condition of spontaneously broken scale symmetry is expressed by

$$\langle 0|\psi^2(x)|0 \rangle \neq 0, \tag{5.161}$$

where we respected the invariance under the field reflection $\psi \rightarrow -\psi$.

The generator for the scale transformation can be obtained from the Nöther current for the scale transformation. An excellent analysis of this was presented in [31]. Here we do not deal with this construction. We state only that the generator can be constructed.

The general discussion in this chapter regarding the spontaneous breakdown of symmetries showed that the main agent of the transformation is the NG boson $\chi(x)$. Its scale transformation together with its translation can be written as

$$\chi(x) \rightarrow e^\theta \chi(e^\theta x) + m\eta(e^\theta - 1) \tag{5.162}$$

with a dimensionless c-number η. The massive quasi-particle field undergoes the following dimensional transformation under the scale transformation:

$$\phi(x, m) \rightarrow e^\theta \phi(e^\theta x, e^{-\theta} m). \tag{5.163}$$

(The generator of the dimensional transformation was given in [32]). Note that the mass m also made the scale change, because the massive free field equation

$$[\partial^\mu \partial_\mu + m^2]\phi(x) = 0 \tag{5.164}$$

should be invariant under this dimensional transformation.

These transformations induce the following change:

$$\left[m + \frac{1}{\eta}\chi(x)\right] \rightarrow e^\theta \left[m + \frac{1}{\eta}\chi(e^\theta x)\right]. \tag{5.165}$$

and

$$\phi(x, \left[m + \frac{1}{\eta}\chi(x)\right]) \rightarrow e^\theta \phi(e^\theta x, \left[m + \frac{1}{\eta}\chi(e^\theta x)\right]). \tag{5.166}$$

Note that as a result of (5.163) m in (5.166) does not change.

We can now prove that the dynamical map of ψ has the form

$$\psi = F[(m + \frac{1}{\eta}\chi), \phi(x, \left[m + \frac{1}{\eta}\chi(x)\right])]. \tag{5.167}$$

This is due to the following reason. Note that both $(m + \frac{1}{\eta}\chi)$ and ϕ are of dimension of inverse of length as ψ is, and that both acquire the same factor e^{θ} under the above transformation. Therefore, the above transformation changes F as

$$F \to [e^{\theta}F]_{x \to e^{\theta}x}. \tag{5.168}$$

This agrees with the behavior of ψ under the scale transformation.

The above expression of the dynamical map can further be simplified by the relation

$$\exp\left(\frac{1}{\eta}\chi(x)\frac{\partial}{\partial m}\right)\phi(x, m) = \phi(x, \left[m + \frac{1}{\eta}\chi(x)\right]). \tag{5.169}$$

In this way [32] the mass can be consistently created in models without any parameters of dimension of length. The scale change of m is induced by the χ-field transformation. We again find the powerful role of the c-q transmutation condition.

It is obvious that the value of m is arbitrary. This is not a problem, because m determines the unit of length.

5.7 REFERENCES

[1] J. Goldstone,. *Nuovo Cimento* **19** (1961) 154.

[2] Y. Nambu and G. Jona Lasinio, *Phys. Rev.* **122** (1961) 345.

[3] J. Goldstone, A. Salam and S. Weinberg, *Phs. Rev.* **127** (1962) 965.

[4] J. C. Ward, *Phys. Rev.* **78** (1950) 182.

[5] Y. Takahashi, *Nuovo Cimento* **6** (1957) 370.

[6] H. Umezawa, *Nuovo Cimento* **40** (1965) 450.

[7] L. Leplae, R. N. Sen and H. Umezawa, *Suppl. Prog. Theor. Phys.*, Communication issue for the 30th Anniversary of the Meson Theory by Dr. H. Yukawa, (1965) 637.

[8] H. Umezawa, H. Matsumoto and M. Tachiki, *Thermo Field Dynamics and Condensed States*, (North-Holland, Amsterdam, 1982).

[9] S. L. Adler, *Phys. Rev.* **B137** (1965) 1022.

[10] F. J. Dyson, *Phys. Rev.* **102** (1956) 1217.

[11] E. Nöther, *Goett. Nachr.* (1918) page 235.

[12] P. C. Hohenberg, *Phys. Rev.* **158** (1967) 383.

[13] N. D. Mermin and H. Wagner, *Phys. Rev. Lett.* **17** (1966) 1133.

[14] S. Coleman, *Comm. Math. Phys.* **3** (1973) 259.

[15] L. Leplae, H. Umezawa and F. Mancini, *Physics Reports* **10C** (1974) 153.

[16] H. Matsumoto, N. J. Papastamatiou and H. Umezawa, *Phys. Rev.* **D12** (1975) 1836.

[17] J. M. Kosterlitz and D. J. Thouless, *J. Phys.* **C16** (1973) 1181.

[18] P. W. Anderson, *Phys. Rev.* **110** (1958) 827.

[19] P. Higgs, *Phys. Rev.* **45** (1960) 1156.

[20] T. W. B. Kibble, *Phys. Rev.* **155** (1967) 1554.

[21] J. Bardeen, L. Cooper and J. Schrieffer, *Phys. Rev.* **108** (1957) 1175.

[22] M. Shah, H. Umezawa and G. Vitiello, *Phys. Rev.* **B10** (1974) 4724.

[23] E. Inonu and E. P. Wigner, *NAS* **39** (1953) 510.

[24] H. Joos and E. Weimar, *Nuovo Cimento* **32** (1976) 283.

[25] E. Weimar, *Acta. Phys. Austr.* **48** (1978) 201.

[26] Y. Fujimoto and N. J. Papastamatiou, *Nuovo Cimento* **40** (1977) 468.

[27] M. Wadati, H. Matsumoto, Y. Takahashi and H. Umezawa, *Fortschr. Phys.* **26** (1978) 357.

[28] M. Wadati, H. Matsumoto and H. Umezawa, *Phys. Rev.* **B18** (1978) 4077.

[29] M. Wadati, *Phys. Rep.* **50** (1979) 87.

[30] U. Krause, J. P. Kuska and R. Wedell, Monovacancy formation energies in cubic crystals, *Preprint* (1989).

[31] C. G. Callan, S. Coleman and R. Jackiw, *Ann. Phys.* **59** (1970) 42.

[32] A. Auliria, N. J. Papastamatiou, Y. Takahashi and H. Umezawa, *Phys. Rev.* **D5** (1972) 3066.

6

Macroscopic Objects

6.1 Boson Transformation

6.1.1 INTRODUCTION

In the last chapter we studied how particle condensations create a rich variety of macroscopic orders in a system of quantum fields. However, in reality we rarely find a pure ordered system; almost all the ordered states carry some kind of defects which have macroscopic structure and distort the orders. Examples are crystal dislocations, magnetic domains in a magnetic order, vortices in superfluid or superconducting order, etc. It is natural to assume then that the origin of these macroscopic objects may be found in the ordered state. Furthermore since the most universal aspect of ordered states is the existence of the NG particles, it is reasonable to expect that these defects are a result of a condensation of the NG particles.

This story suggests a view which is not restricted to ordered state. Not only the NG particles but any boson can form a condensate in a vacuum to create a variety of forms of macroscopic objects. This opens the subject of macroscopic objects created in quantum field systems.

6.1.2 THE PLANCK'S CONSTANT

It is frequently stated that macroscopic or classical phenomena require a vanishing Planck's constant $\hbar$. This view is based on the misconception that quantum means micro and classical means macro. We have seen in the preceding chapters that this is not true.

Indeed, Niels Bohr emphasized that the classical behavior is established when the quantum fluctuation $\hbar\Delta n$ is much smaller than the average quantum number $\hbar n$, i.e. $\Delta n/n \ll 1$. This condition does not depend on $\hbar$, but is controlled by the number of particles participating in the condensation. This shows that the creation of a macroscopic object does not need vanishing $\hbar$ but can use a condensation of a large number of particles [1].

In this book we study macroscopic objects of a quantum field origin in the framework of operator formalism. For an approach based on the functional method readers are advised to see text books such as [2] and [3]. As for the operator formalism for macroscopic objects, see [4, 5].

6.1.3 BOSON TRANSFORMATION

Consider a scalar field ψ satisfying the Heisenberg field equation

$$\Lambda(\partial)\psi(\vec{x},t) = j[\psi](\vec{x},t). \tag{6.1}$$

To distinguish the spatial coordinate from time in this section we denote coordinates by $(\vec{x},t)$. Following the ideas of renormalization theory, we choose Λ such that the free field equation for the quasi-particle field is given by

$$\Lambda(\partial)\varphi(\vec{x},t) = 0. \tag{6.2}$$

Then the equation (6.1) can be put in the integral form

$$\psi(\vec{x},t) = \varphi(\vec{x},t) + \Lambda^{-1}(\partial)j[\psi](\vec{x},t), \tag{6.3}$$

where Λ^{-1} is the Green function for the free field equation (6.2). Feeding this expression repeatedly for $\psi(\vec{x},t)$ on the right hand side of this integral equation, we obtain an expression for ψ in terms of powers of φ. This leads us to one choice for the dynamical map:

$$\psi(\vec{x},t) = F[\vec{x},t:\varphi]. \tag{6.4}$$

Now recall the boson transformation defined in 5.1.4. Apply this to φ:

$$\varphi(\vec{x},t) \rightarrow \varphi(\vec{x},t) + f(\vec{x},t) \tag{6.5}$$

with a c - number function f satisfying the same free field equation:

$$\Lambda(\partial)f(\vec{x},t) = 0. \tag{6.6}$$

It is then obvious from (6.3) that the following operator also satisfies the Heisenberg equation (6.1):

$$\psi(\vec{x},t) = \varphi(\vec{x},t) + f(\vec{x},t) + \Lambda^{-1}(\partial)j[\psi](\vec{x},t), \tag{6.7}$$

which gives a new dynamical map for *the same field* ψ [6]:

$$\psi(\vec{x},t) = F[\vec{x},t:\varphi+f]. \tag{6.8}$$

Then we see that the dynamical map (6.8) is obtained from the one in (6.4) by the boson transformation.

The classical behavior of the macroscopic object is manifested in the vacuum expectation value of ψ:

$$\phi(\vec{x},t) = \langle 0(f)|\psi(\vec{x},t)|0(f)\rangle, \tag{6.9}$$

where $|0(f)\rangle$ is the vacuum with the space-time inhomogeneous boson condensate creating the macroscopic object. The other matrix elements of ψ

describe the behavior of quasiparticles coexisting with the macroscopic object. The quantity $\phi(\vec{x}, t)$ is called the field order parameter. It is due to the presence of space-time dependence in the c-number function f that the field order parameter becomes a function of $(\vec{x}, t)$:

$$\phi(\vec{x}, t) = \langle 0(f)|F[\vec{x}, t : \varphi + f]|0(f)\rangle \qquad (6.10)$$

Taking the vacuum expectation value of both sides of the Heisenberg equation (6.1), we obtain the classical equation for the macroscopic object:

$$\Lambda(\partial)\phi(\vec{x}, t) = < 0(f)|j[\psi](\vec{x}, t)|0(f) > . \qquad (6.11)$$

It is important to note that this classical equation contains effects of Planck's constant $\hbar$. To understand this consider the operator ψ^n. The dynamical map of each ψ contains, not only the c-number ϕ, but also the normal products of quasi-particle operators. Thus, ψ^n contains not only ϕ^n, but also the product of normal product terms. Rearranging a product of normal product terms into a normal product creates c-numbers due to the contraction of creation and annihilation operators of quasi-particles. Note that each contraction creates a c-number of order of $\hbar$. Thus we have

$$\langle 0(f)|\psi^n|0(f)\rangle = \phi^n + 0(\hbar), \qquad (6.12)$$

where $0(\hbar)$ stands for terms of order of positive power of $\hbar$. Thus, only when the contractions which produce the loops in Feynman diagrams are ignored, we can replace $\langle 0(f)|\psi^n|0(f)\rangle$ by ϕ^n. This approximation is called the tree approximation.

Only in the tree approximation does the equation (6.11) take the same form as the Heisenberg equation:

$$\Lambda(\partial)\phi(\vec{x}, t) = j[\phi](\vec{x}, t). \qquad (6.13)$$

Without the tree approximation the classical equation does not have the form of the original Heisenberg equation. Thus the classical limit of quantum field theory does not reproduce the Euler equations for the original Lagrangian. This is remarkable when we consider the fact that, as was pointed out above, loop corrections are of order of the Planck's constant $\hbar$. We have seen in the last subsection that the macroscopic behavior does not mean vanishing $\hbar$ but is created by a condensation of a large number of particles. Therefore, it is not surprising that *the loop corrections do influence the macroscopic behavior* in spite of the fact that they are of order of $\hbar$.

A well known example of modification of the Maxwell equations due to a loop correction in quantum electrodynamics is the modification of the Coulomb potential which caused the Lamb shifts. These are changes in electron energy levels in atoms due to radiative correction.

Historically, quantum field theory was derived from classical field theory by the method of canonical quantization. Now we are deriving classical equations from quantum field theory. We find it very interesting that the exact form of the classical equations, including loop corrections, is not the same as the original classical equations. The original classical equations emerge only in the tree approximation.

Let us summarize the above boson transformation theorem in general terms. Consider a set of Heisenberg fields $\psi_i; i = 1, 2, \cdots, n$ whose dynamics is controlled by certain Heisenberg equations. Consider a solution expressed in terms of quasi-particle fields $\varphi_a; a = 1, 2, \cdots, m$. Notes that m does not need to be equal to n because there can be some composite particles. (Recall that a model of proton and neutron fields with the Yukawa interaction has a solution with three kinds of quasi-particles; proton, neutron and deuteron with the deutron being composite.) This is possible because a finite multiple of countably infinite sets form also a countably infinite set. (The presence of the deuteron does not increase the number of degrees of freedom.) The free field equations for the quasi-particle fields are written as

$$\Lambda_a(\partial)\varphi_a(\vec{x}, t) = 0. \tag{6.14}$$

The dynamical map of the Heisenberg fields are

$$\psi_i(\vec{x}, t) = F_i[\vec{x}, t : \varphi_1, \cdots, \varphi_m]. \tag{6.15}$$

Now perform the boson transformation

$$\varphi_a(\vec{x}, t) \rightarrow \varphi_a(\vec{x}, t) + f_a(\vec{x}, t) \tag{6.16}$$

with

$$\Lambda_a(\partial)f_a(\vec{x}, t) = 0. \tag{6.17}$$

The theorem states that the following dynamical map also satisfies the same Heisenberg equation:

$$\psi_i(\vec{x}, t) = F_i[\vec{x}, t : \varphi_1 + f_1, \cdots, \varphi_m + f_m] \tag{6.18}$$

[1, 7]. The classical fields are then given by the vacuum expectation value of ψ_i.

To simplify notation, we proceed with one Heisenberg field ψ and one quasi-particle field φ in this chapter unless otherwise stated.

6.1.4   INDUCED POTENTIAL AND ZERO ENERGY MODE

The consideration in the last subsection showed that when a macroscopic object is created by boson condensation, the dynamical map of the Heisenberg field ψ has the form $\psi = \phi + \cdots$, in which the dots stand for normal products of operators of quasi-particles. The question now is what kinds of quasi-particles do we have? We will find that macroscopic objects carry certain potentials which act on quantum particles and such potentials create many energy levels for quasi-particles including a zero energy level [8].

For simplicity in this subsection we use the tree approximation, where field contractions are ignored (zeroth order in $\hbar$); the argument can be generalized to all orders of $\hbar$. In this approximation the normal product expansion in any Heisenberg operator in terms of quasi-particle fields is simply an expansion in powers of quasi-particle operators. We thus write the dynamical map of ψ as

$$\psi = \sum_{n=-1}^{\infty} \psi_n, \tag{6.19}$$

where ψ_n is a term with the $(n+1)$-th power of operators of quasi-particles. It is convenient to put this in the form

$$\psi \longrightarrow \psi_\lambda = \sum_{n=-1}^{\infty} \lambda^n \psi_n \tag{6.20}$$

and simultaneously in the Heisenberg equation

$$j[\psi] \longrightarrow \frac{1}{\lambda} j[\lambda \psi_\lambda] \tag{6.21}$$

with $\lambda = 1$. The λ is useful in counting power of operators of quasi-particles, because a λ^n-term contains the $n+1$-th power of operators of quasi-particles.

Now, substitute (6.20) in the Heisenberg equation

$$\Lambda(\partial)\psi = j[\psi] \tag{6.22}$$

and pick up λ^n-terms for every n. We then obtain a hierarchy of equations:

$$\Lambda(\partial)\psi_{-1} = j[\psi_{-1}] \tag{6.23}$$

$$(\Lambda(\partial) - j_1[\psi_{-1}])\psi_0 = 0 \tag{6.24}$$

$$(\Lambda(\partial) - j_1[\psi_{-1}])\psi_1 = \frac{1}{2}j_2[\psi_{-1}][\psi_0]^2 \tag{6.25}$$

etc. In these expressions, $j_n[\phi] \equiv \delta^n j[\phi]/\delta\phi^n$.

Eq.(6.23) shows that ψ_{-1} obeys the classical equation of motion in the tree approximation: it is the order parameter, i.e. $\psi_{-1} = \phi$. From (6.24) we see that ψ_0 obeys a *linear homogeneous* equation, albeit with a potential reflecting the presence of a macroscopic object. Therefore ψ_0 is the quasi-particle field. Eq.(6.25) shows that ψ_1 (and all the subsequent ψ_n) can be expressed in terms of ψ_{-1} and ψ_0 .*Therefore (6.20) with $\lambda = 1$ is the dynamical map in the presence of the macroscopic object.*

Let us simplify the situation by assuming that the macroscopic object under consideration is static. This means that the order parameter $\phi = \psi_{-1}$ does not depend on time. Then, the equation for the quasi-particle field ψ_0 becomes an eigenvalue equation where $j_1[\psi_{-1}]$ acts as potential:

$$\Lambda(-i\omega, \vec{\nabla})u_\omega(\vec{x}) - j_1[\psi_{-1}]u_\omega(\vec{x}) = 0 \qquad (6.26)$$

This equation always has zero eigenvalues.

Indeed, operation of $\vec{\nabla}$ on both sides of the equation (6.23) for static order parameter gives

$$\Lambda(0, \vec{\nabla})\vec{\nabla}\psi_{-1} - j_1[\psi_{-1}]\vec{\nabla}\psi_{-1} = 0 \qquad (6.27)$$

This shows that (6.26) with $\omega = 0$ has the solution $\vec{\nabla}\psi_{-1}$.

Thus the quasi-particle field ψ_0 contains as its part the term of form of $(1/\sqrt{2\omega})(\vec{\chi} + \vec{\chi}^\dagger)\vec{\nabla}\phi$, where χ is the annihilation operator of the zero energy mode and ω is its energy. This expression is however not possible, because ω is zero. This reflects the fact that, since the discrete zero energy mode is not observed as an excitation mode, it is not a particle mode. Since we cannot make use of the operators $(\vec{\chi}, \vec{\chi}^\dagger)$, we change the operators to the canonical operators

$$\vec{q} = \frac{1}{\sqrt{2\omega}}(\vec{\chi} + \vec{\chi}^\dagger) \qquad (6.28)$$

$$\vec{p} = -i\sqrt{\frac{\omega}{2}}(\vec{\chi} - \vec{\chi}^\dagger), \qquad (6.29)$$

which give

$$[q_i, p_j] = i\delta_{ij}. \qquad (6.30)$$

We proceed to formulate our argument in terms of $\vec{q}$ and $\vec{p}$ in order to avoid appearance of zero energy in the denominator.

Therefore the quasi-particle field ψ_0 has the form

$$\psi_0 = -\vec{q} \cdot \vec{\nabla}\psi_{-1} + \varphi, \qquad (6.31)$$

where φ is the quasi-particle field in which the zero energy mode is omitted [9]. The annihilation operator of this quasi-particle field φ is denoted by α_k. The wave functions of α_k in φ are eigenfunctions with non-zero eigenvalues and, therefore, are orthogonal to $-\nabla_i\psi_{-1}$. This means that the *quantum coordinate* $\vec{q}$ and its conjugate momentum $\vec{p}$ commute with α_k. In this way we can treat this zero energy mode separately from other particle levels, because $\vec{q}$ acts as a quantum mechanical operator rather than a quantum field operator.

Now the dynamical map (6.20) reads as

$$\psi = \phi - \vec{q} \cdot \vec{\nabla}\phi + \varphi + \cdots, \qquad (6.32)$$

where dots stand for higher order products consisting of normal products of φ-field together with $\vec{q}$ and $\vec{p}$. We can rewrite this as

$$\psi(x) = \phi(\vec{x} - \vec{q}) + \varphi + \cdots. \qquad (6.33)$$

Note that $\vec{x}$ and $\vec{q}$ appear through the combination $(\vec{x} - \vec{q})$.

In this subsection we learned of two features of macroscopic objects created by particle condensation. First, the macroscopic objects carry potentials which influence wave functions of quasi-particles. This potential is called the self-consistent potential or macroscopic potential. Second, a zero energy mode appears. This is the NG mode associated with spontaneously broken space translation symmetry caused by the presence of a macroscopic object; since the position of the macroscopic object is arbitrary, its shift does not consume energy, giving rise to the zero energy mode. This creates the quantum mechanical degrees of freedom associated with $(\vec{q}, \vec{p})$.

We have formed the picture that a condensation of quasi-particles creates a new vacuum with a macroscopic object, and then this macroscopic object carries a macroscopic potential which acts on the states of quasi-particles. A well known example of this macroscopic potential is the potential made by vortices in a mixed state in superconductivity; the vortex potential is approximately given by the pair field order parameter $\Delta(\vec{x}) = \langle 0(f)|\psi_\downarrow \psi_\uparrow|0(f)\rangle$ which acts on electrons. The energy carried by electrons trapped by this potential is called the core energy, which has a significant contribution to the free energy of mixed states.

6.1.5 QUANTUM MECHANICAL OPERATORS

In the last subsection we learned how quantum mechanical operators $\vec{q}$ and $\vec{p}$ emerge when a macroscopic object is created in a quantum field system. We saw that c-number position $\vec{x}$ and q-number position $\vec{q}$ appeared through the combination $(\vec{x} - \vec{q})$. In this subsection we show that this appearance of $\vec{q}$ is due to the c-q transmutation rule.

To see this, recall the c-q transmutation rule discussed in section 1.4 and in section 5.5. In particular, in 5.5 it was shown that the $\vec{x}$-dependence of order parameters for crystals appears only through the combination $[\vec{x} + (1/c)\vec{\chi}]$ with the phonon field $\vec{\chi}$. The same argument is applied to the order parameter $\phi(\vec{x}, t)$. Thus, in the dynamical maps of Heisenberg operators, $\vec{x}$ appears always through the combination $\vec{x} - \vec{q}$ in such a manner that the q-number transformation $\vec{q} \rightarrow \vec{q} - \vec{\theta}$ induces the c-number transformation $\vec{x} \rightarrow \vec{x} + \vec{\theta}$ as $\vec{x} - (\vec{q} - \vec{\theta}) = (\vec{x} + \vec{\theta}) - \vec{q}$ [8, 1].

Since $\vec{x}$ is real, $\vec{q}$ is Hermitian. The generator of this $\vec{q}$-translation should be the momentum operator $\vec{p}$ canonically conjugate to $\vec{q}$, because the latter generates the translation of $\vec{q}$. This should also be Hermitian. Furthermore, invariance of the Lagrangian under the spatial translation dictates that a constant shift of the origin of the spatial coordinates should not induce any physical change and should not consume any energy, implying that this origin shift is a zero energy mode. This shows that $\vec{p}$ is independent of time [8, 1]:

$$\dot{p}_i = -i[p_i, H] = 0. \tag{6.34}$$

There is a significant difference between the phonon field χ in crystals and the operator $\vec{q}$ considered here. The phonons are particle modes with gapless energy; their lowest energies are zero. This is because the phonon modes are caused by lattice oscillation. On the other hand, when presence of macroscopic objects spontaneously breaks spatial symmetry, there is no reason for having particle NG modes, because the NG mode in such a case reflects the fact that the position of the macroscopic object is arbitrary; shifting the position of the object does not consume energy. Thus this NG mode is a discrete zero-energy mode. In other words, the zero energy mode associated with $\vec{p}$ is the NG mode which is a result of spontaneous breakdown of spatial translation symmetry; the presence of spatially inhomogeneous macroscopic objects makes the vacuum $|0(f)\rangle$ vary under the spatial translation, inducing the NG zero energy mode.

As we have seen, $\vec{q}$ and $\vec{p}$ form a *quantum mechanical* set of canonical operators:

$$[q_i, p_j] = i\delta_{ij}. \tag{6.35}$$

Exercise 1 Discuss the c-q transmutation of spatial rotation.

Hint: Consider a rotation

$$x_i \rightarrow x_i' = c_{ij}x_j, \quad c_{ik}c_{jk} = \delta_{ij}. \tag{6.36}$$

Perform the q-number transformations

$$q_i \rightarrow q_k c_{ki}, \quad p_i \rightarrow p_k c_{ki}. \tag{6.37}$$

This induces the transformation:

$$x_i - q_i \rightarrow (x_k' - q_k)c_{ki}. \tag{6.38}$$

This has two effects; it rotates all the q-number vectors and it makes the replacement $\vec{x} \rightarrow \vec{x}'$. The generator for this is the angular momentum operator

$$\vec{L}[q, p] = \vec{q} \times \vec{p}. \tag{6.39}$$

However, this is not sufficient; the annihilation operator of quasi-particle α_k carries the momentum suffix, which should also be rotated as $\vec{k} \rightarrow \vec{k}'$. As is well known, this is generated by the usual angular momentum operator of quantum fields:

$$l_i = -\frac{i}{2}\epsilon_{ijl} \int d^3k\, \alpha_k^\dagger \left(k_j \frac{\partial}{\partial k_l}\right) \alpha_k + h.c. \tag{6.40}$$

where $h.c.$ means the Hermitian conjugate of the preceding term. Thus the quasi-particle operators, which do not contribute to the total momentum

$\vec{P}$, do contribute to the angular momentum which is the generator of the spatial rotation.

When a macroscopic object is not spherically symmetric, this breaks the rotational symmetry of vacuum. This spontaneously broken rotational symmetry creates a new zero energy NG mode, giving rise to a rotational quantum mechanical operator $\vec{s}$ with the commutation relations

$$[s_i, s_j] = i\epsilon_{ijk}s_k. \tag{6.41}$$

This operator is called the quantum mechanical spin operator.

The generator of rotation is thus given by

$$\vec{L} = \vec{L}[q, p] + \vec{l} + \vec{s}. \tag{6.42}$$

End of Exercise

Since a rotation of coordinates should not change the energy, it commutes with the Hamiltonian: $[H, \vec{L}] = 0$. This implies that $\vec{L}$ is independent of time:

$$\dot{L}_i = 0. \tag{6.43}$$

In this book, for simplicity, we do not consider the cases with $\vec{s}$, unless otherwise stated. Thus

$$\vec{L} = \vec{L}[q, p] + \vec{l}. \tag{6.44}$$

The vacuum expectation value of this is the quantum mechanical part:

$$\langle 0(f)|\vec{L}|0(f)\rangle = \vec{L}[q, p]. \tag{6.45}$$

The Hilbert space is the product of the quasi-particle Fock space and the quantum mechanical Hilbert space. The quasi-particle Fock space is built by cyclic operation of creation operators $\alpha_k^\dagger$ on the vacuum $|0(f)\rangle$ which contains the particle condensation characterized by the boson transformation function $f(\vec{x}, t)$. This Fock space is denoted by $\mathcal{H}(f)$ while the quantum mechanical Hilbert space associated with $(\vec{q}, \vec{p})$ is written as $\mathcal{H}(q, p)$. The entire state vector space is $\mathcal{H}(f) \times \mathcal{H}(q, p)$.

In this subsection we saw that a system of quantum fields gives rise to all kinds of degrees of freedom; α_k for quasi-particles, $(\vec{q}, \vec{p})$ for quantum mechanical degrees of freedom and $\phi(\vec{x})$ for classically behaving macroscopic objects. In the next subsection we will show that this argument is based on the asymptotic condition, as is the quasi-particle picture in quantum field theory without any macroscopic object.

6.1.6   ASYMPTOTIC CONDITION FOR FREE PARTICLES

In this section we assume that the field order parameter ϕ in (6.9) damps fast enough at the infinitely far domain $|\vec{x}| \to \infty$ so that it is Fourier transformable. Then, the quantum particles far from the domain of macroscopic objects become free; we have again the situation in which the asymptotic fields are free due to the same reason as the one for the asymptotic field discussed in subsection 4.6.3. We may use these free fields for the free quasi-particles as we did in subsection 4.6.3.

Since the creation of these free particles is commutable with spatial translation, the quantum mechanical operators commute with the quasi-particle operators:

$$[\vec{q}, \alpha_k] = [\vec{p}, \alpha_k] = 0. \qquad (6.46)$$

as was pointed out in the last subsection. (The role of quantum mechanical operators is similar to the one for the collective coordinates [10] common in the literature based on functional integral formalism. A relation between the quantum coordinate and the collective coordinate was discussed in [9].)

The complete form of the dynamical map has the form

$$\psi(\vec{x}, t) = F[\vec{x} - \vec{q}, p : \alpha_k^\dagger, \alpha_k]. \qquad (6.47)$$

Recall (6.34), which means that the dynamical map of H does not contain $\vec{q}$.

This last property of the Hamiltonian is an example of a general feature of global operators, namely those operators which are spatial integrations of certain local operators, as indicated in section 4.10. Since $\vec{q}$ appears always together with $\vec{x}$ as $\vec{x} - \vec{q}$ in any local operator, its spatial integration eliminates $\vec{q}$. We thus see that the dynamical map of global operators, say N, cannot contain $\vec{q}$:

$$N(t) = N[t : \vec{p}, \alpha_k^\dagger, \alpha_k]. \qquad (6.48)$$

Furthermore, in section 4.6 where quantum field theory without any macroscopic objects was discussed, we showed that the dynamical map of the Hamiltonian contains the quasi-particle operators only through the number operator

$$N_k = \alpha_k^\dagger \alpha_k. \qquad (6.49)$$

The same argument can be applied to the case under consideration. Thus we obtain a dynamical map of the form

$$H = H[\vec{p} : N_k]. \qquad (6.50)$$

As was shown in (6.39), the generator of rotation, $\vec{L}$, contains $\vec{q}$ as $\vec{q} \times \vec{p}$. This is because $\vec{L}$ is not a global operator. To see this recall the angular momentum operator in a scalar quantum field theory:

$$L_i = \epsilon_{ijk} \int d^3x \, x_j T_{k0}, \qquad (6.51)$$

where $T_{\mu\nu}$ is the energy stress tensor. Since this integrand contains the pure c-number x_j explicitly, L_i is not a global operator.

We close this subsection with two very important remarks:

The first is that the vacuum expectation value and other matrix elements among state vectors in the Fock space $\mathcal{H}(f)$ are not c-numbers but quantum mechanical operators consisting of $\vec{q}$ and $\vec{p}$. Under appropriate conditions it is often possible to construct wavepackets with reasonably sharp values of position and velocity. In such cases, one may crudely replace $\vec{q}$ and $\dot{\vec{q}}$ by their expectation values $\vec{c}$ and $\vec{v}$. The extended object is then called classical. *In this classical limit (6.11) becomes a purely classical (i.e. c-number) equation.* In any case, in order to obtain c-numbers we must consider also the matrix elements among state vectors in $\mathcal{H}(q, p)$.

The second remark is the following: although the field order parameter ϕ is a functional of the boson transformation function f, the requirement that ϕ should be Fourier transformable *does not prohibit singular boson transformation functions f* which do not have a Fourier transform. As a matter of fact, singular boson transformation functions play a very significant role in the physics of macroscopic objects created by particle condensation.

6.1.7 Momentum Operator

We now introduce an important theorem which states that the macroscopic objects determine the spatial coordinate system. This originates from the fact that the position of the asymptotic domain in which the quantum particles become free is defined in relation to the position of macroscopic objects. In order for the particles to be free, they must be far from the location of the objects (see subsection 4.6.3).

As has been pointed out, *the quasi-particle fields, and hence the full Heisenberg field, must depend on $\vec{x}$ only through the combination $\vec{x} - \vec{q}$:*

$$\psi(\vec{x}, t) = F[\vec{x} - \vec{q}, \vec{p} : \varphi(\vec{x} - \vec{q}, t, \vec{p})]. \tag{6.52}$$

This means that the dynamical map of the total momentum operator which generates the spatial translation does not contain quasi-particle operators:

$$\vec{P} = \vec{p}. \tag{6.53}$$

6.1.8 Quasi-Static Objects and Singularities

When an object can be brought to rest in a particular coordinate system it is called quasi-static.

A theorem of central significance is the statement that *the boson transformation parameter $f(\vec{x}, t)$ for a quasi-static object is not Fourier transformable unless the quasi-particle energy ω_k vanishes for certain values of $\vec{k}$.* Consider a quasi-static object and make use of the coordinate system in which it is at rest. Then, its f does not depend on time: $f = f(\vec{x})$. Assume first that this f has a Fourier transform and write its Fourier amplitude as $f(\vec{k})$. Then, the free field equation (6.6) becomes $\omega_k f(\vec{k}) = 0$. This gives $f(\vec{k}) = 0$ for $\omega_k \neq 0$. The theorem immediately follows from this.

When ω_k vanishes at a nonvanishing value of $\vec{k}$, say $\vec{k}_1$, $f(\vec{k}_1)$ does not need to vanish. Then there can be a static object with spatial periodicity $2\pi/|\vec{k}_1|$. Many spin sinusoidal phases in metals are examples of this kind.

In the following we do not consider objects of this kind. We thus assume that $\omega_k \neq 0$ for $\vec{k} \neq 0$. We also ignore the case of constant f, which corresponds to a spatially homogeneous ground state.

Then the above theorem states that $f(x)$ for a static object is not Fourier transformable. Since the boosting operation moves the object from its rest system to a moving system, we find that the $f(\vec{x}, t)$ for a quasi-static object carries a singularity which prohibits its Fourier transform. We are going to see that these singularities are of a topological nature. These objects are called *topological objects*. Note that, even though f has no Fourier transform, the field order parameter $\phi(\vec{x}, t)$ is Fourier transformable. In the following part of this chapter we confine our attention to quasi-static objects.

As we have seen in the preceding subsections in this chapter, the boson transformation function depends on $\vec{x}$ only through the combination $\vec{x} - \vec{q}$. Thus, this function for a macroscopic object at rest has the form

$$f = f(\vec{x} - \vec{q}). \tag{6.54}$$

Suppose that the Lagrangian is invariant under a certain form of boosting, which changes the rest system to a moving system with certain velocity. Denote the generator for this boosting by $\vec{K}$. This transforms the q-number $\vec{q}$ into $\vec{q}'$. In this boosted coordinate $(\vec{x}, t)$ and $\vec{q}$ should appear in a combination in which the transformation of $\vec{q}$ induces the boosting of $\vec{x}$ (the c-q transformation rule). This combination defines the generalized coordinate $\vec{X}$ [11, 12, 1]:

$$\vec{X} = \vec{X}(\vec{x} - \vec{q}, t, \vec{p} : \theta). \tag{6.55}$$

Here θ is a parameter which specifies the position and velocity of the macroscopic object.

The expectation values of $\vec{q}$ and $\vec{p}$ with respect to the state vectors in the quantum mechanical Hilbert space $\mathcal{H}(q, p)$ determines the position and velocity vectors of the macroscopic object.

In these moving coordinates the boson transformation function is

$$f = f(\vec{X}). \tag{6.56}$$

Since the free field equation (6.6) for f should be invariant under this boosting, this boosted f is also a solution of the same free field equation.

Although the boosting generator $\vec{K}$ is given by spatial integration of a local density, this local density is not a pure operator but contains the c-numbers $\vec{x}$ and t explicitly:

$$K_i = \int d^3x [x_i T_{00}(x) - t T_{i0}(x)]. \tag{6.57}$$

Thus, $\vec{K}$ is not a global operator and, therefore, its dynamical map may contain $\vec{q}$.

6.1.9 MANY OBJECTS AND MANY CONSERVATION LAWS

The boson transformation is a nice way of linearizing the nonlinear effects in field order parameters for macroscopic objects. This becomes clear when we treat many macroscopic objects. They are simply obtained from a boson transformation parameter which is a *linear* superposition of the boson transformation parameters of single objects, because the equation (6.6) for f is linear and homogeneous.

In the last subsection we considered a boosted object created by a boson transformation function $f(\vec{X}(\vec{x} - \vec{q}, t, \vec{p} : \theta))$. This with any choice for θ is a solution of (6.6); θ specifies the position and velocity of the object. Therefore, a linear sum of f with n-choices of θ creates n objects:

$$f_n(x) = \sum_{i=1}^{n} f(\vec{X}(\vec{x} - \vec{q}_i, t, \vec{p}_i : \theta_i)). \tag{6.58}$$

The θ_i specify the position and velocity of the i-th quasi-static object. Since the $\vec{x}$ in any f_i can be shifted independently of the $\vec{x}$ in the other f_i, each f_i carries its own quantum mechanical operators $(\vec{q}_i, \vec{p}_i)$; we have n mutually commuting sets of quantum mechanical operators. The $\vec{q}_i$ is called the position operator of the i-th object, while $\vec{p}_i$ is called the momentum operator.

Using this f and following the method in section 6.1.3, we can treat a set of many macroscopic objects [12]. The total momentum should generate a translation of $\vec{x}$ in all the f_i simultaneously. It is given by

$$\vec{P} = \vec{p} = \sum_i \vec{p}_i. \tag{6.59}$$

As was discussed in section 6.1.7, there is no contribution from quantum particles to $\vec{P}$.

The situation is different in the case of other transformations, because they transform the quasi-particles too. However, when we consider the vacuum expectation values of these generators, there is no contribution from quasi particles and we find purely quantum mechanical operators. Then, there are generators of these transformations acting on each object individually, and the generator for the whole system is the sum of these. Thus, denoting the total Hamiltonian, the angular momentum and the boosting generator by H, $\vec{L}$ and $\vec{K}$ respectively, we have

$$\langle 0(f)|H|0(f)\rangle = \sum_i H_i[p_i], \tag{6.60}$$

$$\langle 0(f)|\vec{L}|0(f)\rangle = \sum_i \vec{L}_i[q_i, p_i] \tag{6.61}$$

$$\langle 0(f)|\vec{K}|0(f)\rangle \ = \ \sum_i \vec{K}_i[q_i,p_i]. \qquad (6.62)$$

These are quantum mechanical operators.

In quantum field theory all conserved quantities are associated with the Nöther currents (see section 5.1.6.). The Nöther currents associated with the time and space translational invariances lead to the total Hamiltonian and total momentum respectively. Thus a question arises, how can so many momenta $\vec{p}_i$ and Hamiltonians H_i appear. The answer is the following. The total energy of a system of many objects is a function of the velocities of these objects: $E(\vec{v}_1, \vec{v}_2, \vec{v}_3, \ldots)$. Note that the velocity parameters which are functions of parameters $\vec{\theta}_i$ do not appear in the original field equations, but only in the solutions. Then we have new conserved quantities given by $\partial E/\partial v_i$. *This illustrates how the single Nöether conservation law creates many conservation laws in the multi-object sector* when there exists more than one macroscopic object.

6.1.10 EMERGENT SYMMETRIES

In the last subsection we saw an example in which a system with a Lagrangian invariant under certain space-time transformations, consisting of time and space translations, rotation and boosting, manifests in observations multiples of the same space-time symmetries. This illustrates those cases in which observed symmetries contain new symmetries which do not exist in the basic Lagrangian. This phenomenon is called emergent symmetry.

The reason for the appearance of emergent symmetries lies in the fact that the observed symmetries come from the structure of the dynamical map of Heisenberg operators. Since the dynamical maps can contain many parameters θ_i (such as the velocity of a macroscopic object), a change in these parameters brings the vacuum through a set of energy degenerate vacua, giving rise to new symmetries.

The point of significance is that the creation of macroscopic objects may induce emergent symmetries, depending on the structure of the objects. The lattice structure formed by vortices in superconductors is an example of this kind [1].

6.2 Relativistic Models

6.2.1 (1+1)-DIMENSIONAL OBJECTS

Since ultimately the quantum mechanical degrees of freedom are present to implement the space - time invariant transformations of the macroscopic object, a study of how these transformations are effected sheds much light on their properties. We illustrate this for the case of a system in one space dimension with a relativistically invariant equation of motion. The Poincaré algebra is

$$[H, p] = 0 \qquad (6.63)$$

$$[K, H] = -ip \qquad (6.64)$$

$$[K, p] = -iH \qquad (6.65)$$

Let us begin with the dynamical map of the Hamiltonian H. When we consider the rest system of the macroscopic object ($p = 0$), the energy is a sum of the mass of the object and energy of free quasi-particles:

$$M(N_k) = M_0 + H_0, \qquad (6.66)$$

This is the rest energy operator. Here M_0 is a c - number called the mass of the macroscopic object and H_0 is the free Hamiltonian of the quasi particles:

$$H_0 = \int dk\, \omega_k \alpha_k^\dagger \alpha_k. \qquad (6.67)$$

Since the space-time coordinate is attached to the macroscopic object, $M(N_k)$ is invariant under the Poincaré group transformation.

Furthermore, according to (6.50), the dynamical map of the Hamiltonian of the system consists of the momentum p and the quasi-particle number $N_k = \alpha_k^\dagger \alpha_k$: $H = H[p, N_k]$. Then special relativity tells us that the dynamical map of H is [11]

$$H(p : n_k) = \sqrt{p^2 + M^2(N_k)}. \qquad (6.68)$$

Note that M contains neither q nor p but N_k only. This means that the frequently written relation $H = H_0 + \sqrt{p^2 + M_0^2}$ is not correct in the rigorous sense.

The above result for H shows that

$$\dot{q} = -i[q, H] = \frac{p}{\sqrt{p^2 + M^2(N_k)}}. \qquad (6.69)$$

Now the equations (6.64) and (6.69) determine the dynamical map of the boost generator as

$$K = -\frac{1}{2}(qH + Hq) \qquad (6.70)$$

This can be inverted:

$$q = -\frac{1}{2}(KH^{-1} + H^{-1}K). \qquad (6.71)$$

The relation (6.69) gives

$$p = \frac{M\dot{q}}{\sqrt{1 - \dot{q}^2}}. \qquad (6.72)$$

This shows that $\dot{q}$ acts as the velocity under Lorentz boosting.

Now it is easy to find the generalized coordinate introduced in (6.55). Since $\dot{q}$ is the boosting velocity, we should have

$$X(x - q, t, p) = \frac{1}{2}\left[(x - q)\frac{1}{\sqrt{1 - \dot{q}^2}} + \frac{1}{\sqrt{1 - \dot{q}^2}}(x - q)\right]. \quad (6.73)$$

(See [11, 12, 1] for a more rigorous derivation of this generalized coordinate.) Note the symmetrization in (6.73) which makes X Hermitian.

The generalized time coordinate $T(x - q, t, p)$ for scalar fields is determined by the requirement that the transformation $(x, t) \rightarrow (X, T)$ keeps the Dalembertian invariant:

$$\partial_t^2 - \partial_x^2 = \partial_T^2 - \partial_X^2 \quad (6.74)$$

together with the requirement that $T(x, t : 0) = t$. These requirements lead to

$$T(x - q, t, p) = (1 - \dot{q}^2)^{1/2}t - \frac{1}{2}\dot{q}[(x - q)\frac{1}{(1 - \dot{q}^2)^{1/2}} + \frac{1}{(1 - \dot{q}^2)^{1/2}}(x - q)]. \quad (6.75)$$

In the history of the development of quantum mechanics there were several attempts to quantize systems with Hamiltonians of the form $\sqrt{p^2 + M_0^2}$. However, these attempts were not successful and this led Dirac to propose his famous relativistic wave equation, opening the way to quantum field theory. In the framework of relativistic wave equations, Newton and Wigner introduced an operator which describes the position of a relativistic object [13, 14]. Now, starting with quantum field theory, we obtain macroscopic objects with Hamiltonians of the form $\sqrt{P^2 + M^2}$ and the quantum mechanical operator q. A remarkable fact [11, 15] is that *the expression in (6.71) for q is exactly the same as the one for the Newton-Wigner position operator.*

When there are n macroscopic objects, we have n sets of the operators: (q_i, p_i). We then have

$$< 0(f)|H|0(f) > = \sum_i H_i, \quad (6.76)$$

$$P = \sum_i \vec{p}_i, \quad (6.77)$$

$$< 0(f)|K|0(f) > = \sum_i K_i. \quad (6.78)$$

where $H_i = \sqrt{p_i^2 + M_0^2}$ and the $\vec{K}_i$ are given by expressions analogous to (6.70):

$$K_i = -\frac{1}{2}(q_i H_i + H_i q_i) \quad (6.79)$$

This shows emergent n-fold Poincaré symmetries in a system with a basic Lagrangian which has a single Poincaré symmetry.

6.2.2   THREE DIMENSIONAL MODELS

The consideration of the last subsection on $(1+1)$-dimensional relativistic macroscopic objects can be extended to $(1+3)$-dimensional relativistic macroscopic objects. Since this extension requires a complex and lengthy consideration, we here list the main results, leaving details to [15]. Since this section has not much relation to other parts of this book, readers may skip this subsection.

We consider a single object in a relativistic model of three spatial dimensions. The suffix signifies the space vector component: e.g. $i = 1, 2, 3$.

The main part of the algebraic relations for the Poincaré algebra is

$$i[K_i, P_j] = H\delta_{ij}, \tag{6.80}$$
$$i[K_i, H] = P_i, \tag{6.81}$$
$$i[K_i, K_j] = \epsilon_{ijk}L_k, \tag{6.82}$$
$$[L_i, L_j] = i\epsilon_{ijk}L_k. \tag{6.83}$$

Here, H, $\vec{P}$, $\vec{K}$ and $\vec{L}$ are the Hamiltonian, momentum, boosting operator and angular momentum, respectively.

We list the dynamical maps of these operators:

$$H = [p^2 + M^2]^{1/2}, \tag{6.84}$$
$$\vec{P} = \vec{p}, \tag{6.85}$$
$$\vec{L} = \vec{q} \times \vec{p} + \vec{l}, \tag{6.86}$$
$$\vec{K} = -\frac{1}{2}(H\vec{q} + \vec{q}H) - \frac{\vec{p} \times \vec{l}}{H + M}. \tag{6.87}$$

Here M has the structure in (6.66):

$$M = M_0 + H_0 \tag{6.88}$$

with M_0 being the mass of the macroscopic object and H_0 is the free Hamiltonian of the quasi-particles:

$$H_0 = \int dk\,\omega_k \alpha_k^\dagger \alpha_k. \tag{6.89}$$

The $\vec{l}$ is the contribution of quasi-particles to the angular momentum as was shown in (6.40).

The quantum mechanical coordinate $\vec{q}$ is expressed in terms of these operators as

$$\vec{q} = -\frac{1}{2}\left(H^{-1}\vec{K} + \vec{K}H^{-1}\right) - \frac{\vec{p} \times \vec{l}}{H(H + M)}. \tag{6.90}$$

This is the Newton-Wigner position operator in three dimensions. When the object is not spherically symmetric so that it breaks the rotational symmetry, $\vec{l}$ should be replaced by $\vec{l} + \vec{s}$ where $\vec{s}$ is the zero energy spin operator explained in section 6.1.5.

The generalized coordinate in tree approximation is (see p. 432 in [1])

$$X_i(x - q, t, p) = \eta_{ij}(p)(x - q)_j, \tag{6.91}$$

where

$$\eta_{ij}(p) = \delta_{ij} + \frac{p_i p_j}{M(H + M)}. \tag{6.92}$$

The generalized time coordinate is

$$T(x - q, t, p) = \frac{H}{M}t - \frac{1}{M}\vec{p} \cdot (\vec{x} - \vec{\xi}), \tag{6.93}$$

where

$$\vec{\xi} = \vec{q} - \frac{\vec{p} \times \vec{l}}{M(H + M)} + \frac{i}{2}\vec{p}H^{-2}. \tag{6.94}$$

6.2.3   MACROSCOPIC OBJECTS WITH A FERMION

We determined the generalized coordinates for the scalar field by studying Klein - Gordon type equations. There, the generalized coordinates were so made that the space-time transformations of c-numbers $(\vec{x}, t)$ are induced by certain q-number transformations. However, when a spinor field is interacting with a macroscopic object, we need to prove that the dynamical map of the Dirac field Heisenberg operator is also written in terms of the same generalized coordinates. We prove this as follows (see [16]).

Let us define Γ_μ by

$$\gamma_\mu \partial^\mu = \Gamma_\mu D^\mu, \tag{6.95}$$

where D^μ is the derivative vector associated with $(\vec{X}, T)$ and γ_μ are the Dirac matrices. In the following we show that there exists a Hermitian operator which consists of H and $\vec{p}$ and which gives rise to Γ_μ as follows:

$$\Gamma_\mu = A\gamma_\mu A^{-1}. \tag{6.96}$$

This indicates that *the dynamical map of the spinor Heisenberg field Ψ has the form*

$$\Psi(\vec{x}, t) = A\Phi[\vec{X}, T], \tag{6.97}$$

where Φ is a four-component column function of $\vec{X}$ and T, each component being a scalar field. The spinor transformation of Ψ is taken care of by the operator A, which transforms as a spinor. We can generalize this argument to quantum fields of any spin by using the multi-spinor representation for the Lorentz group.

In the following we study the spinor operator A by considering $(1+1)$-dimensional systems. This situation arises, for example, in polyacetylene [17, 18, 19, 20].

In (1+1)-dimensions, the Dirac matrices are two by two matrices. The generalized coordinates (X, T) were given in (6.73) and (6.75) in subsection 6.2.1. With these results for the generalized coordinate, equation (6.95) together with (6.96) is solved to give

$$A = \frac{1}{\sqrt{2M(H + M)}}[H + M + \gamma_5 p]. \tag{6.98}$$

where $\gamma_5 = \gamma_0 \gamma_1$.

Now, (6.97) puts the dynamical map of the electron field in the following form:

$$\Psi(x, t) = A\Phi(X, T). \tag{6.99}$$

The spinor operator A in (6.98) can be rewritten as

$$A = \frac{\sqrt{P_+} + \sqrt{P_-}}{2\sqrt{M}} + \gamma_5 \frac{\sqrt{P_+} - \sqrt{P_-}}{2\sqrt{M}}, \tag{6.100}$$

where $P_{\pm} = H \pm p$. Here use was made of the relations

$$(\sqrt{P_+} + \sqrt{P_-})^2 = 2(H + M), \tag{6.101}$$
$$(\sqrt{P_+} - \sqrt{P_-})^2 = 2p. \tag{6.102}$$

The fact that A transforms as a spinor can be seen from the relations

$$[A, H] = [A, p] = 0, \tag{6.103}$$
$$[A, K] = \frac{i}{2}\gamma_5 A. \tag{6.104}$$

There are several interesting models of interacting fermions and boson fields with a soliton such that there appears a massless fermion quasi-particle [21, 22, 23, 24]. In some of these models the massless fermion is the Goldstino (that is a NG fermion) associated with the breakdown of supersymmetry caused by the soliton.

We assume a model with a fermionic zero-energy mode, the annihilation operator of which is denoted by α:

$$[\alpha, \alpha^\dagger]_+ = 1, \tag{6.105}$$
$$[\alpha, H] = [\alpha, P] = [\alpha, K] = 0. \tag{6.106}$$

Here the suffix + for the square bracket denotes an anticommutator. In deriving these relations we considered the following facts. The α commutes with H because it is the zero energy operator so that it does not contribute to the free Hamiltonian H_0. The velocity of massless particles does not depend on the boosting. Therefore, α commutes with K. It commutes with P because no particle operators contribute to P.

Now introduce the operators

$$S = \sqrt{2M} A\gamma_0 s\alpha \; , \quad \bar{S} = \sqrt{2M} s^\dagger \gamma_0 A\gamma_0 \alpha^\dagger, \qquad (6.107)$$

where s is a two - component c-number spinor which satisfies

$$s = -i\gamma_1 s^*, \quad s_i s_j^\dagger = \frac{1}{2}(1 - i\gamma_1)_{ij}. \qquad (6.108)$$

When use is made of the representation

$$\gamma_0 = \begin{pmatrix} 0 & 1 \\ 1 & 0 \end{pmatrix}, \quad \gamma_1 = \begin{pmatrix} i & 0 \\ 0 & -i \end{pmatrix}, \qquad (6.109)$$

we have

$$s = \begin{pmatrix} 1 \\ 0 \end{pmatrix}, \qquad (6.110)$$

and

$$\frac{1}{2}(1 - i\gamma_1) = \begin{pmatrix} 1 & 0 \\ 0 & 0 \end{pmatrix}. \qquad (6.111)$$

We can then derive

$$[S, p_\mu] = 0, \quad [S, K] = -\frac{i}{2}\gamma_5 S, \qquad (6.112)$$

$$[S, \bar{S}]_+ = \gamma^\mu p_\mu - i\gamma_5 M, \qquad (6.113)$$

where $p_\mu = (H, p)$. Here we used the relation

$$M A\gamma_0 (1 - i\gamma_1)\gamma_0 A\gamma_0 = \gamma^\mu p_\mu - i\gamma_5 M. \qquad (6.114)$$

This means that the operators S and $\bar{S}$ satisfy a supersymmetry algebra with the central charge being given by the soliton mass M.

This appearance of supersymmetry [25] is remarkable in the sense that the existence of such a symmetry is not obvious in structure of the original Lagrangian. This may be another example of emergent symmetry. It has been shown that the fractional fermion number which was first discovered by Jackiw [21] follows from this.

In the case of (1+3)-dimensional systems, it was shown in [16] that the operator A is the four-by-four matrix operator defined by

$$A = \frac{1}{\sqrt{2M(H + M)}}[H + M + \gamma_0\vec{\gamma}\vec{p}]. \qquad (6.115)$$

6.3 Quantum Solitons in Tree Approximation

6.3.1 STABLE SOLITONS

In this subsection we realize the general ideas described in preceeding subsections by studying quasi-static objects in (1+1)-dimensional space. Thus

x and k are not vectors but assume real values. To make our consideration concrete, we limit ourselves to the sine-Gordon model. Thus, the Heisenberg equation is

$$- (\partial_t^2 - \partial_x^2)\psi(x,t) = \frac{\mu^2}{\lambda} \sin [\lambda \psi(x,t)]. \tag{6.116}$$

Note that this equation is relativistically invariant.

The integral form of the Heisenberg equation (6.7) has the form

$$\psi = \varphi + f + (-\partial_t^2 + \partial_x^2 - m^2)^{-1} \left(\frac{\mu^2}{\lambda} \sin [\lambda \psi] - m^2 \psi \right), \tag{6.117}$$

where m is the renormalized mass. The φ is the quasi-particle free field satisfying the Klein-Gordon equation with mass m. In this section we work in the tree approximation, thus ignoring all loop corrections. In this approximation we have $m = \mu$.

The equation (6.6) for the boson transformation parameter is

$$(\partial_t^2 - \partial_x^2 + m^2)f = 0. \tag{6.118}$$

The static solutions are $f = \exp[\pm mx]$, which diverge at $x = \pm\infty$, and hence prohibits the Fourier transformation of f.

The quantum mechanical effects are incorporated by replacing x by the generalized coordinate X. Since the model is relativistically invariant, $X(x - q, t, p)$ is given by (6.73):

$$X(x - q, t, p) = \frac{1}{2} \left[(x - q)\frac{1}{\sqrt{1 - \dot{q}^2}} + \frac{1}{\sqrt{1 - \dot{q}^2}}(x - q) \right]. \tag{6.119}$$

Thus f for a single soliton is given by

$$f(x,t) = f[X] = \exp[mX]. \tag{6.120}$$

We chose the plus sign as a matter of convention.

Taking the vacuum expectation value of both sides of (6.117), we obtain

$$\phi(x,t) = f[X(x - q, t : \dot{q})]$$
$$+(-\partial_t^2 + \partial_x^2 - m^2)^{-1} \left(\frac{\mu^2}{\lambda} \sin \lambda\phi(x,t) - m^2\phi(x,t) \right). \tag{6.121}$$

This gives [26]

$$\phi(x,t) = 4 \tan^{-1} f[X]. \tag{6.122}$$

which is the well-known kink solution of the sine-Gordon equation. Note that this ϕ is Fourier transformable, although f is not so.

The Hamiltonian H, boosting generator K and momentum operator form the (1+1)-dimensional Poincaré algebra [12]. We have $H = \sqrt{p^2 + M^2}$ with

$M = M_0 + H_0$ in agreement with the general considerations of the previous section. According to (6.70) we also have

$$K = -\frac{1}{2}(qH + Hq).$$ (6.123)

When we have many solitons, $f(x,t)$ is the sum of $f[X_i]$ with $X_i = X(x - q_i, t, p_i : \theta_i)$. The Poincaré group generators (H_i, K_i, p_i) are associated to each soliton, and the generators of different solitons commute with each other.

For two solitons the classical equation for the field order parameter is (6.121) with $f[X]$ being replaced by $f[X_1] + f[X_2]$. The solution is [12]

$$\phi(x,t) = 4\tan^{-1}\left|\frac{f(X_1) + f(X_2)}{1 - af(X_1)f(X_2)}\right|,$$ (6.124)

where

$$a = \tanh^2\frac{\theta_1 - \theta_2}{2}$$ (6.125)

with θ_i being defined by $\dot{q}_i = \tanh\theta_i$.

There are other kinds of solutions. An analysis of the so-called breather solution along the lines of this book was presented in [27].

6.3.2   UNSTABLE SOLITONS

When a $(1+1)$-dimensional model has stable solitons, we say that the classical equation is integrable. The sine-Gordon equation discussed previously is an example of this. A significant and well known example of a soliton in condensed matter physics is the one in polyacetylene.

However, most of the integrable equations are derived from the quantum field Heisenberg equation in the tree approximation. The loop corrections can easily make the classical equations none integrable. The inclusion of finite temperature effects or of the effects of impurities also spoils integrability. There are many other causes of a loss of integrability.

For example, the SSH-model [17, 18, 19] for polyacetylene has a lattice structure with lattice length a. In the limit of vanishing a there emerges a continuum model which is called the TLM-model [20]. The TLM model gives an equation which is integrable and therefore has static solitons. Expanding the SSH-Lagrangian in powers of a up to order a^2 , one naturally finds an acoustic phonon interaction effect which makes the new equation none integrable. An approximate treatment of this acoustic phonon interaction shows that the solitons change in time with speed of the order of magnitude of the acoustic phonon velocity [28]. A careful numerical analysis of the SSH-model was reported in [29]. The result shows that for a short initial time a soliton moves with a speed of the order of magnitude of the acoustic phonon velocity, but soon the tail starts to oscillate and spread and finally the soliton dies away.

Thermal effects on solitons along the Josephson junction have been studied by Scott and his colleagues [30, 31].

6.4   Topological Objects and High Dimension

6.4.1   PATH-DEPENDENT TOPOLOGY

In this chapter we have studied macroscopic objects created by quasi-particle condensation, which is expressed through the boson transformation of the quasi-particle field. The soliton in one dimensional space discussed in subsection 6.3 is an example in which the boson transformation function f has a singularity which prohibits the Fourier transform of f. There, the singularity is the kind which makes f exponentially divergent either at $x = \infty$ or at $x = -\infty$. This object is a topological object because, as it was shown in (6.122) its order parameter is multi-valued as a functional of f; ϕ assumes the same value for $f \to f + 2\pi n$ with any integer n.

In one dimensional space we can not create any path-dependent topology. However, in spaces of dimension higher than one we have functions which explicitly depend on a path. They are multi-valued as a function of $\vec{x}$ and are said to be spatially topological. These functions have a certain singular domain which makes the functions multi-valued. A simple example is the cylindrical angle around a line; this line is called the line singularity.

The topological singularity also prohibits the Fourier transform. When the boson transformation function f has a spatially topological singularity, the macroscopic objects are called spatially topological objects [7, 1].

It is intuitively obvious that in a domain of topological singularity

$$[\nabla_i, \nabla_j]f \neq 0 \tag{6.126}$$

for some choices of (i, j).

In case of the cylindrical angle θ around a straight line going through the origin along the third axis, we have

$$\nabla^2\theta = 0, \tag{6.127}$$
$$\vec{\nabla} \times \vec{\nabla}\theta = 2\pi\vec{e}_3\delta(x_1)\delta(x_2), \tag{6.128}$$

where $\vec{e}_3$ is the unit vector along the third axis.

Extending this consideration to space and time, we define geometrically topological objects by the condition

$$[\partial_\mu, \partial_\nu]f \neq 0. \tag{6.129}$$

for some choices of (μ, ν).

A significant theorem is that bosons whose condensation creates geometrically topological objects are energy gapless. This indicates a close relationship between the ordered states and geometrically topological objects.

Any order is maintained by certain NG-bosons, whose energies are gapless. Therefore, condensation of these NG-bosons can create geometrically topological objects. These objects are called defects, because the order is destroyed in the topologically singular domain. Therefore, it is natural that almost all ordered states contain some defects. The nature of the defects is controlled by the original symmetry which is spontaneously broken by the order.

In this section we ignore the quantum mechanical operators. We can revive them by following the consideration of the last section.

6.4.2 SUPERFLUID VORTICES

Let us begin with a simple example by using the relativistic scalar model for superfluidity considered in section 5.2. There, the phase symmetry associated with the phase transformation of a scalar Heisenberg field $\psi(x)$ is spontaneously broken. Since the NG-boson field $\chi(x)$ is massless, we have:

$$\partial^2 \chi = 0. \tag{6.130}$$

The dynamical map of the Nöther current was given in subsection 5.2.9:

$$N_\mu(x) = \frac{1}{c}\partial_\mu\chi(x) + \cdots. \tag{6.131}$$

Here the c-number constant c is such that the boson transformation

$$\chi \rightarrow \chi + cn\theta \tag{6.132}$$

changes the phase of the Heisenberg field ψ with the phase factor $\exp[in\theta]$. We have replaced the parameter θ by $n\theta$ with an integer n for sake of later convenience.

Let us now ask if we can choose the cylindrical angle in (6.127) and (6.128) for the θ in this boson translation.

First, for a static f, the free field equation (6.130) becomes the Laplace equation (6.127), implying that the free field equation for f is satisfied by the cylindrical angle.

Second, the order parameter and any other observed quantities should be single valued. The phase factor of the order parameter $\phi = \langle 0(f)|\psi|0(f)\rangle$ is $\exp[in\theta]$ which has a same value for $\theta \rightarrow \theta + 2\pi$, implying that the order parameter is single valued.

We thus see that θ can be the cylindrical angle.

This boson transformation changes N_i as

$$N_i = \frac{1}{c}\nabla_i\chi + n\nabla_i\theta. \tag{6.133}$$

Here the dots in (6.131) are ignored.

This creates the macroscopic current $J_i = \langle 0(f)|N_i|0(f)\rangle$ as follows:

$$J_i(x) = n\nabla_i\theta = n\frac{1}{r}\vec{e}_i, \quad (i = 1, 2) \tag{6.134}$$

where $\vec{e}_i$ is the unit vector along the i-th direction. This is the well known superfluid vortex. Since the topological singularity lies on the third axis, we have $J_3 = 0$.

The above result gives

$$\oint d\vec{s} \cdot \vec{J} = 2\pi n. \tag{6.135}$$

In this case the circular integration is around the third axis, giving us the so-called the superfluid vortex flux quantization.

This relation can be put in the form of a surface integration

$$N_T = \int_S dS\vec{e}_3 \cdot (\vec{\nabla} \times \vec{J}) = 2\pi n, \tag{6.136}$$

where S is the surface area spanning in the contour in (6.135).

The N_T is referred to as the topological charge, which is always quantized, and called a *macroscopic quantum number*.

In the last chapter we saw that many ordered states are macroscopic quantum states. Now we see that some quantum numbers are also macroscopic. The gap between micro and macro has now become very narrow.

Here are some notes of significance. The phase ordered state is a result of breakdown of symmetry associated with the phase transformation which is Abelian. Condensation of NG-boson creates a vortex which also manifests an Abelian symmetry, because the cylindrical angle is a parameter of rotation which forms an Abelian group. We thus find a remarkable mechanism; the defect in an ordered state created by spontaneous breakdown of certain symmetry breaks the order and manifests the original symmetry in the coordinate space.

Another note is the following. In condensed matter physics there is a physical requirement which states that the current should be well defined everywhere. This implies that the vortex current should vanish at the center which is the topologically singular point, since intuitively, circular current has no sense at the center. In the above model the current does not vanish at the center; it diverges at the center. This is due to the relativistic nature of the model. The superfluid models in condensed matter physics are not relativistic and the vortex current carries a form factor which eliminates the current at the center. This will be seen in the next subsection when we study the superconducting vortex.

In condensed matter physics we find that the smallest topological charge $n = 1$ is frequently preferred, because the vortex contribution to the free energy is usually proportional to n^2, in favor of smaller n.

6.4.3 SUPERCONDUCTING VORTICES

The study of superconducting current can follow the same steps as used in the last subsection for superfluid vortex. We use the BCS-model discussed in subsection 5.3.2.

As it was shown in (5.114) the dynamical map of the electric current has the form:

$$\vec{j} = \frac{1}{4\pi\lambda_L^2}c(\nabla^2)[\vec{a} - \frac{1}{e}\eta^{-1}(\nabla^2)\vec{\nabla}\chi] + \cdots \tag{6.137}$$

with the condition

$$c(0) = 1. \tag{6.138}$$

The constant λ_L is called the *London penetration length*. The $c(\nabla^2)$ in (6.137) is the form factor of the Meissner current and is called the boson characteristic function.

On the other hand, our study in section 5.3.4 showed that the dynamical map of the electron Heisenberg fields $\psi_{\uparrow,\downarrow}$ has the phase factor $\exp[i(1/\eta(\nabla^2))\chi]$, where χ is the NG-boson.

Thus the electron phase factor $\exp[i(n/2)\theta]$ is generated by

$$\chi(x) \rightarrow \chi(x) + n\frac{1}{2}\eta(\nabla^2)\theta \tag{6.139}$$

with an integer n.

Here we need to explain why we consider an electron phase change not of θ but by $(1/2)\theta$. Since the order parameter ϕ is $\langle 0(f)|\psi_{\uparrow}\psi_{\downarrow}|0(f)\rangle$, this phase change creates the phase factor $\exp[in\theta]$ in the order parameter. Furthermore, since the electron field is a fermion field, any observable quantities contain an even number of the electron fields. Thus, the single-valued condition is not applied to the electron field but to the order parameter. This requires that not $\exp[in(1/2)\theta]$ but $\exp[in\theta]$ should be single valued.

When we ignore the high energy longitudinal plasma, χ satisfies the Laplace equation.

$$\nabla^2\chi = 0. \tag{6.140}$$

Let us note that the Fourier transformable solution of the Laplace equation (6.140) picks up only the zero momentum contribution. We saw in the last chapter that the Coulomb interaction brings up the Nambu - Goldstone mode to the plasma mode for the entire momentum domain *except at zero momentum*. Therefore, only the observable boson excitation levels are the plasma levels; the gapless excitation levels (the phason levels) are eliminated from observation by the Coulomb effect. This is called the Anderson-Higgs-Kibble mechanism. However, the phase field satisfying the Laplace equation (6.140) cannot be ignored, because its presence is required by the phase invariance. The phase transformation of the electron field is generated by the translation of the χ, which requires the zero momentum and zero energy part of the NG field. Though this boson is not observed as an excitation level, this is the agent for the phase transformation. Furthermore, as we saw in the last subsection, topologically singular solutions of the Laplace equation can create vortices.

Following the consideration of the last subsection, we choose the cylindrical angle θ around the third axis for the θ in the above boson transformation. The dynamical map (6.137) for the current shows that this boson transformation creates a macroscopic current [7, 1]. Then the Maxwell equations dictate that a classical vector potential $\vec{A}$ is induced. Since we are making use of the Coulomb gauge, this is a transverse vector; $\vec{\nabla} \cdot \vec{A} = 0$. The gauge invariance tells us that the macroscopic current which follows from the dynamical map (6.137) (with the terms denoted by dots ignored) is

$$\vec{J} = \frac{1}{4\pi\lambda_L^2}c(\nabla^2)[\vec{A} - n\frac{1}{2e}\vec{\nabla}\theta]. \tag{6.141}$$

Note that the cylindrical angle θ is an electron phase, but *it is not a gauge function*, because θ is not Fourier transformable. Indeed, although $\vec{\nabla}\theta$ would look like a longitudinal quantity, the fact that $\vec{\nabla} \times \vec{\nabla}\theta \neq 0$ shows that it does contribute to the electromagnetic field.

The first term in (6.141) is the Meissner current which makes superconductors diamagnetic. The transverse vector potential $\vec{A}$ is determined by the Maxwell equations which in static cases read as

$$\nabla^2\vec{A} = -4\pi\vec{J}. \tag{6.142}$$

(Note that in this subsection we use Gaussian units.)

Feeding the current given by (6.141) into this equation, we obtain

$$\nabla^2\vec{A} = -\frac{1}{\lambda_L^2}c(\nabla^2)[\vec{A} - n\frac{1}{2e}\vec{\nabla}\theta]. \tag{6.143}$$

Operating with $(\vec{\nabla}\times)$ on both sides of this, we obtain the following equation for the magnetic field $\vec{H}$:

$$\nabla^2\vec{H} = \frac{1}{\lambda_L^2}c(\nabla^2)[\vec{H} - n\phi\vec{e}_3\delta(\vec{x})], \tag{6.144}$$

where use was made of the relation (6.128). Here $\phi = (\pi/e)$ is the constant called the unit flux. The $\vec{x}$ is the two dimensional vector in the $(x_1 x_2)$-plane.

This gives $\vec{H}(\vec{x}) = H(\vec{x})\vec{e}_3$ with

$$H(\vec{x}) = \frac{n\phi}{\lambda_L^2} \int \frac{d^2k}{(2\pi)^2} \frac{c(-k^2)}{k^2 + c(-k^2)/\lambda_L^2} e^{i\vec{k}\vec{x}}. \tag{6.145}$$

Here $c(-k^2)$ is the Fourier form for $c(\nabla^2)$.

Integrating this over the $(x_1 x_2)$-plane, we obtain

$$\Phi = \int d^2x H(\vec{x}) = n\phi. \tag{6.146}$$

Since this is an integral multiple of ϕ, it is called the flux quantization, which is another example of a macroscopic quantum number. The name "unit flux" for ϕ originates from this flux quantization.

The presence of the Meissner term in the current makes the vortex current short ranged; the $1/r$ in the superfluid current is replaced by $[\exp(-r/\lambda_L)]/r$ when we crudely put $c(-k^2) = 1$ for the Meissner form factor. This form factor is important, because $c(-k^2)$ with high k damps so fast that the vortex current vanishes at the center. (See the requirement for vortex center discussed in the last subsection.)

6.4.4   A GENERAL FORMALISM

We saw in previous sections some examples of macroscopic objects created by boson transformation functions f which are not singlevalued. In this subsection we briefly sketch the general formalism for geometrical topological objects.

A systematic procedure for constructing topologically singular f starts by introducing an auxiliary tensor $G_{\mu\nu}^{\dagger} = [\partial_\mu, \partial_\nu]f$. Clearly $G_{\mu\nu}^{\dagger}(\vec{x}, t)$ has support only in domains where f has topological singularities. The existence of a path-dependent f requires that ∂f be single valued: $[\partial_\mu, \partial_\nu]\partial_\rho f = 0$.

A macroscopic object created by the boson transformation with a topologically singular f is called a topological object. By choosing the domain of singularity to be a line, closed surface, etc. we can create topological objects in the form of vortices, bags, etc.

If the mass of a quasi-particle is m, the equation (6.6) can be put in the form

$$[-\partial^2 + m^2]f(\vec{x}, t) = 0, \tag{6.147}$$

where $\partial^2 \equiv \partial^\mu \partial_\mu$. Then we can derive

$$\partial^\mu G_{\mu\nu}^{\dagger}(\vec{x}, t) = [-\partial^2 + m^2]\partial_\nu f(\vec{x}, t). \tag{6.148}$$

Therefore we assume that $\partial_\mu f$ and $G_{\mu\nu}^{\dagger}$ are Fourier transformable, and we have

$$\partial_\mu f = \frac{1}{-\partial^2 + m^2}\partial^\lambda G_{\lambda\mu}^{\dagger}. \tag{6.149}$$

Since $G_{\mu\nu}^{\dagger}$ is antisymmetric, this implies that $\partial^2 f = 0$, which in its turn shows $m = 0$.

Thus we obtain the remarkable result that *topological objects in spaces of two or more dimensions can be created only by the condensation of gapless energy bosons.*

If $G_{\mu\nu}^{\dagger}$ is given, ∂f is computed from (6.149) with $m = 0$. This determines a path-dependent f. This in turn fixes $G_{\mu\nu}^{\dagger}$, which should agree with the original $G_{\mu\nu}^{\dagger}$.

This is a self-consistency condition for $G^\dagger_{\mu\nu}$. This condition is automatically satisfied if instead of $G^\dagger_{\mu\nu}$ we start from its dual

$$G^{\mu\nu} = -\frac{1}{2}\epsilon^{\mu\nu\lambda\rho}G^\dagger_{\lambda\rho}, \tag{6.150}$$

where $\epsilon^{\mu\nu\lambda\rho}$ is the totally antisymmetric tensor with the property $\epsilon^{0123} = 1$. With (6.148) this leads immediately to the continuity equation $\partial_\mu G^{\mu\nu} = 0$. It has been shown that this continuity relation is all that is needed to satisfy the consistency condition mentioned above [1].

Thus *a study of topological objects begins with solving this continuity equation* to obtain $G^{\mu\nu}$, from which we obtain $G^\dagger_{\mu\nu}$ by means of (6.150); this in turn determines ∂f and therefore a path-dependent f.

For example, suppose that we want f to be singular along a line parameterized by a spatial parameter σ together with a time parameter τ as $y_\mu(\tau, \sigma)$. Since $G^{\mu\nu}(x, t)$ should have support only on this line, it is given by

$$G^{\mu\nu}(x, t) = c \int d\tau \int d\sigma \frac{\partial[y_\mu, y_\nu]}{\partial[\tau, \sigma]} \delta^{(4)}[x - y(\tau, \sigma)], \tag{6.151}$$

with a constant c and the Jacobian $\partial[y_\mu, y_\nu]/\partial[\tau, \sigma]$:

$$\frac{\partial[y_\mu, y_\nu]}{\partial[\tau, \sigma]} = \frac{\partial y_\mu}{\partial\tau}\frac{\partial y_\nu}{\partial\sigma} - \frac{\partial y_\mu}{\partial\sigma}\frac{\partial y_\nu}{\partial\tau}. \tag{6.152}$$

We can simplify the example by choosing y_0 for τ:

$$y_0(\tau, \sigma) = \tau. \tag{6.153}$$

Denote the end points of the line at time τ by $\sigma_1(\tau)$ and $\sigma_2(\tau)$. Then, we find

$$\partial_\mu G^{\mu\nu}(x) = c \int d\tau \frac{\partial y_\nu}{\partial\tau} \delta^{(4)}[x - y(\tau, \sigma)]|^{\sigma_1(\tau)}_{\sigma_2(\tau)}. \tag{6.154}$$

This gives

$$\begin{aligned}\partial_\mu G^{\mu\nu}(x) &= c\delta[\vec{x} - \vec{y}(t_x, \sigma_1(t_x))]\dot{y}_\nu(t_x, \sigma_1(t_x)) \\ &\quad -c\delta[\vec{x} - \vec{y}(t_x, \sigma_2(t_x))]\dot{y}_\nu(t_x, \sigma_2(t_x)). \end{aligned} \tag{6.155}$$

Therefore, the continuity equation requires that the line of singularity should not have end points. Note that the end points at the boundary surface of a system are not the end points of the singularities because the boundary surface itself is a topological surface singularity.

The constant c is related to the topological charge defined by

$$N_T = \frac{1}{2}\int_S dS^{\mu\nu} G^\dagger_{\mu\nu}. \tag{6.156}$$

The topological charge is conserved along the line y. An example of such a charge is the quantum flux for the superfluid vortex and magnetic flux for the superconducting vortex current. When we make a complete circuit around the line, all of the observable quantities should come back to their original values because they should be single-valued. Typically this condition leads to the quantization of the topological charge.

When the line y is straight, f is proportional to the cylindrical angle θ around this line. Then this general formalism reduces to the consideration of the previous subsections.

6.4.5 DEFECTS AND THE ORIGINAL SYMMETRY

Since the phase field in superconductivity satisfies the Laplace equation, so does the boson transformation parameter f in $\chi \to \chi + f$. Our analysis of topological objects in the previous subsection showed that this transformation creates an assembly of topological line singularities which are the vortices. These objects are called the defects in a superconductor, because the order is distorted in a domain around the singular line.

As mentioned above the straight vortex line is created by $f = \theta$ with θ being the cylindrical angle. We observe that the phase parameter θ in $\exp[i\theta]$ and the cylindrical angle θ are both Abelian transformation parameters. This leads to a remarkable phenomenon. The superconducting state is a phase - ordered state which is created by spontaneous breakdown of phase symmetry. This phase ordered state is maintained by the phase boson. The defects are the results of condensation of these phase bosons and manifest the original Abelian symmetry through their spatial structure.

In general terms this phenomenon can be summarized as follows. Consider an ordered state created by the spontaneous breakdown of certain symmetry. This ordered state is maintained by the NG-bosons. Condensation of these bosons creates certain defects, which manifest the original symmetry through their space time structure.

In this section, we considered mostly the defects associated with some Abelian order. A well known one of the same kind is the crystal defect. Since the crystal order is a result of spontaneous breakdown of spatial translation symmetry, which is Abelian, the defects are objects with line topological singularities. In crystal there are three NG-bosons (phonons) forming a vector. Therefore, the topological charge is also a vector with three components. The crystal defects with line topological singularities are called dislocations and their topological charge vector is the Burgers vector. The Burgers vector is also quantized. All of the crystal defects are aggregates of line singularities. The point defects are small spherical surfaces made by assembly of enclosed lines. A systematic analysis of point defects in many crystals was made on the basis of the formalism given in the last subsection and numerical results of point defect energy were compared with experiments [32].

In ordered states with spontaneously broken non Abelian symmetries too, condensation of the NG-bosons creates macroscopic objects. A general formalism for such macroscopic objects is much more complicated than the one for Abelian type. A version of this kind of formalism was presented in [1]. The non Abelian type permits a rich variety of macroscopic objects including points such as monopoles [33]. When we extend our argument to include space and time topological objects, the variety still increases. The instanton is a well known example (e.g., see [2]).

However, real situations can be still more complex. This is due to the emergent symmetries discussed in subsection 6.1.10. Depending on structure of macroscopic objects, creation of these objects can create a variety of new orders. For example, in mixed states of superconductors, vortices form a lattice. The lattice length of this vortex lattice is not a parameter present in the basic Lagrangian; this parameter appears only in a particular state among many permitted by the dynamics. (Recall the soliton velocities in multi soliton systems; appearance of these velocity parameters creates multi Poincaré symmetries). This provides quantum field theory with a rich capability for describing a variety of natural phenomena. However, it also makes it very difficult to exhaust all of the possible states when a basic Lagrangian is given.

An interesting case is the problem of a high density assembly of topological objects. A simple amorphous material made by parallel line singularities is a relatively easy object. It is known that conservation of the Burgers vectors along the line tends to make such a material harder to cut. However, more random distribution of macroscopic objects may require a complex method of analysis.

Since these macroscopic objects deform ordered states, they may frequently create complicated curved geometry. Papini made an attempt to derive a curved geometry on basis of the general formalism for topological objects presented in the last subsection [34].

The study of macroscopic objects of quantum origin is a rich and complex subject which has just started, and we have had only a glimpse of the subject.

6.5   Finite or Infinite Volume

6.5.1   Two Views

We have seen many magical effects caused by the infinite number of degrees of freedom in quantum field theory. Some readers might then say that, since every object has a finite size, the number of degrees of freedom is finite. This statement would have been correct in the view of the old atomism in which the observed particles and basic entities are the same, but it is no longer the case in modern many body systems described by quantum field theory. Even with a finite volume a quantum field system carries an infinite number of degrees of freedom.

Still we might feel that the origin of the appearance of many inequivalent Fock spaces is due to the infinite volume, and therefore, they would not appear in reality at least in condensed matter physics because most of the samples we find in condensed matter physics are of finite size. Indeed, appearance of inequivalent Fock spaces is usually caused by the presence of two state vectors which differ from each other by an infinite particle number. Assuming a spatially homogeneous system, in order for any condensation of particles to have an observable effect, its density should be finite so that the infinite volume makes total particle number infinite. This argument might suggest that infinite volume is the origin of inequivalent Fock spaces.

This argument has a flaw, however because a finite size has a boundary which always breaks spatial translation symmetry, which was the basic assumption in the above argument.

When we meet a system with a finite size, there usually are two different views. One is to quantize the field in this finite volume, assuming that this is the entire world. This view is obviously not rigorously correct unless the boundary surface is a rigid surface which acts as an infinitely high repulsive potential separating perfectly the outside from the inside. To be realistic, such a boundary is unlikely. This leads us to another view, in which we treat the boundary itself as a part of a system which lies in our world of infinite volume. In this view we quantize fields in a world of infinite volume and look for a solution of basic equations for the Heisenberg fields such that the solution self-consistently contains a boundary surface as a macroscopic object.

The two views have an essential difference, because a self-consistently created boundary surface may carry a variety of degrees of freedom such as infinite choices of its shape, the surface vibration creating surface phonons, surface spin oscillation which creates surface magnons, and so on. By including these degrees of freedom, the system carries a sufficient number of degrees of freedom to create inequivalent Fock spaces.

6.5.2   CONDENSED MATTER PHYSICS

To make our idea concrete, let us consider a superconducting system of a finite size. The situation inside the domain far from the surface is described approximately by the solution in a system of infinite size. Since the phase symmetry is broken, the total charge carried by electrons appears to be different from the one in the normal conducting state. Certainly, such a charge deviation is impossible to occur; the apparent deviation is accumulated in the domain around the surface. The situation in the domain not far from the boundary surface is not the same as the one deep inside the domain, and the state around the domain frequently assimilates the normal phase, though it gradually changes into superconducting phase when we move into the inside domain.

Furthermore, as was stated previously, the boundary surface itself may carry phonons, magnons and other quantum particles.

Since the surface itself is a part of our object, it provides us with a target of research called surface physics.

The boundary surface as a macroscopic object may also create a self-consistent potential acting on quantum particles, as it was discussed in the case of solitons.

Strictly speaking, the presence of the boundary surface creates quantum mechanical operators which we have studied in previous sections. However, usually the size is large enough for these operators to be disregarded.

6.5.3 NUCLEAR PHYSICS AND HIGH ENERGY PHYSICS

We can apply a similar view to large nuclei. This means that some number of nucleons self-consistently create a finite domain to which the nucleons are confined. The boundary surface naturally gives rise to surface vibration modes, surface rotation modes, etc.

Furthermore, the surface objects naturally create a self-consistent potential, which traps nucleons. Crudely approximating this surface by a sphere, this potential is a function of the radius. Expanding this in powers of the radius and ignoring terms of higher order than the second, the potential becomes approximately the oscillator potential. In this way we may have a simple shell model. We need to improve the approximation to have a realistic model.

We have seen that macroscopic objects frequently create new symmetries. These were called the emergent symmetries. A nonspherical shape may create rotational modes and quadrupole moments. These freedoms could create the Elliot type symmetry group which consists of one harmonic oscillator, three rotational generators and five quadrupole operators of traceless second rank symmetric tensor, forming U(3)-algebra [35]. There the quantum mechanical operators studied in this chapter might play an important role.

This line of thought suggests a tempting challenge of deriving the celebrated interacting boson model [36] from quantum field models [37, 38].

The small nuclei such as the deuteron are usually treated as composite particles instead of extended objects. It is not clear how this simple composite particle picture and the picture of nucleons trapped by macroscopic objects are related to each other.

A similar view arises also at much more microscopic level, that is, the problems of particles consisting of quarks; composite model, bag model, etc.

6.6   Biological Order

When we have a general consistent formalism for macroscopic orders, it is tempting to try to apply it to highly ordered systems. This naturally brings our attention to order in biological systems. We may easily guess the significance of electric effects in biological systems. In his pioneering paper [39] Fröhlich proposed a model of dipolar quantum oscillation along one-dimensional chains of protein molecules. A similar mechanism with soliton waves was proposed by Davidov [40]. These dipolar quantum modes seem to appear in most biological objects and is stimulating experimental studies. This line of research has recently been pursued actively by Del Giudice, Vitiello and their colleagues [41, 42].

Among the many biological objects a particularly interesting one is the brain. For any theory to be able to claim itself as a brain theory, it should be able to explain the origin of such fascinating properties as the mechanism for creation and recollection of memories and consciousness.

For many years it was believed that brain function is controlled solely by the classical neuron system which provides the pathway for neural impulses. This is frequently called the neuron doctrine. However, more and more experimental data seem to be contradicting this neuron doctrine. The most essential one among many facts is the nonlocality of memory function discovered by Pribram. On this and other subjects of the brain, readers are advised to read a book written by Pribram himself [43].

There have been many models based on quantum theories, but many of them are rather philosophically oriented. The article by Burns in [44] provides a detailed list of papers on the subject of consciousness, including quantum models. The incorrect perception that the quantum system has only microscopic manifestations considerably confused this subject. As we have seen in preceding sections, manifestation of macroscopic ordered states is of quantum origin. When we recall that almost all of the macroscopic ordered states are the result of quantum field theory, it seems natural to assume that macroscopic ordered states in biological systems are also created by a similar mechanism. It is reasonable to assume that memories in the brain manifest certain orders which are associated with some ordered states. Then it is worthwhile to try the assumption that these ordered states are controlled by the same mechanism as the one for the ordered states explained in this book.

Let us ask what are the immediate results, when we assume that ordered states in the brain are indeed the macroscopic quantum ordered states described in this book [45, 46]. The resulting intuitive picture is that certain external stimuli print some ordered pattern on the brain system through a condensation mechanism. In this view the brain is regarded as an ordered vacuum. Since we have huge choices for these kinds of ordered vacua, it is not surprising that memory in the brain has huge varieties. Furthermore, these memories are relatively stable because they appear as vacuum states.

There are more remarkable results which immediately follow our assumption. Since these ordered states are maintained by the gapless energy NG modes, even very weak external stimuli can easily excite the NG modes. These NG excitations then remind us that we do have a memory associated with the orders maintained by these NG modes. As soon as the NG excitation disappears, we are no longer conscious of the presence of the memory. Thus the recollection mechanism is an immediate consequence of the theory. We have learned that these orders created in a system of quantum fields have a nonlocal manifestation. Thus, this kind of theory does not support the neuron doctrine, but agrees with Pribram's conclusion.

In this view the classical neuron system simply plays the role of a pathway for the conduction of neural impulses which may be important in the transmission of signals, but memory-related functions are controlled by a much wider domain inside and outside of neuron systems. This domain is a large one without any sharp boundary. To develop such a theory we need some degrees of freedom which play the role analogous to that of electrons in metals. Since these degrees of freedom are some quantum fields residing in the cortex, we may call them corticons [46]. We need also other boson degrees of freedom to act like photons in metals.

This is a rather abstract presentation for a brain theory when the view of macroscopic quantum ordered states for brain systems is adopted. However, to make it a real scientific theory, we should know what is a corticon and what are the bosons interacting with the corticons.

Recently, such a concrete theory for the brain was presented in [47] and it is called the quantum brain dynamics (QBD). In this paper the authors build a concrete theory of the kind mentioned above on the basis of the present knowledge of the brain's structure. This theory thus considers both the extra and intracellular regions which consist of a three dimensional network of microscopic protein filaments. These filaments carry the Frölich type dipolar quanta. This is the concrete identification of the corticons. The bosons interacting with the corticons are assumed to be the phonons associated with the water molecules which form deformable quasi crystals due to hydrogen bonds. The nonvanishing order parameter is the electric dipole formed by a systematic arrangement of electric dipoles. It is not surprising that this ordered system of brain turns out to be the one proposed by Del Giudice, Vitiello et al for a general living system. This kind of analysis may result in a mathematical theory applicable to many biological phenomena and in this way they can be tested as theories for the brain.

A remarkable fact is that, as was pointed out above, this structure of protein filament networks is not restricted to the brain but is universal to most biological cells. Through this structure, the entire body as a whole may form a system of intensive correlations [47].

6.7 REFERENCES

[1] H. Umezawa, H. Matsumoto, and M. Tachiki, *Thermo Field Dynamics and Condensed States*, (North-Holland, Amsterdam, 1982).

[2] R. Rajaraman, *Solitons and Instantons*, (North Holland, Amsterdam, 1982).

[3] P. J. Caudrey, *Solitons*, (Springer Verlag, Berlin, 1980).

[4] N. J. Papastamatiou and H. Umezawa, In W. Rozmus and J. A. Tuszynski, editors, *Non Linear and Chaotic Phenomena in Plasma Solids and Fliuds-Proceeding of CAP-NSERC Summer Institute*, (World Scientific, Singapore, 1991) 97.

[5] R. Manka and G. Vitiello, *Ann. Phys.* **61** (1990) 199.

[6] H. Matsumoto, P. Sodano and H. Umezawa, *Phys. Rev.* **D19** (1979) 511.

[7] L. Leplae, H. Umezawa and F. Mancini, *Physics Reports* **10C** (1974) 153.

[8] H. Matsumoto, G. Oberlechner, M. Umezawa and H. Umezawa, *J. Math. Phys.* **20** (1979) 2088.

[9] G. Semenoff, H. Matsumoto and H. Umezawa, *Prog. Theor. Phys.* **67** (1982) 1619.

[10] J. L. Gervais, A. Jevicki and B. Sakita, *Phys. Rev.* **D12** (1975) 1038.

[11] H. Matsumoto, N. J. Papastamatiou, G. Semenoff and H. Umezawa, *Phys. Rev.* **D24** (1981) 406.

[12] H. Matsumoto, H. Umezawa and N. J. Papastamatiou, *Phys. Rev.* **D28** (1983) 1434.

[13] T. D. Newton and E. P. Wigner, *Rev. Mod. Phys.* **21** (1949) 400.

[14] T. F. Jordan, *J. Math. Phys.* **21** (1980) 2028.

[15] H. Yamanaka, H. Matsumoto and H. Umezawa, *Phys. Rev.* **D24** (1981) 2607.

[16] H. Matsumoto, G. Semenoff and H. Umezawa, *Prog. Theor. Phys.* **69** (1983) 1631.

[17] W. P Su, J. R. Schrieffer and A. J. Heeger, *Phys. Rev.* **B22** (1980) 2099.

[18] W. P Su, J. R. Schrieffer and A. J. Heeger, *Phys. Rev.* **B28** (1983) 1138.

[19] W. P Su, J. R. Schrieffer and A. J. Heeger, *Phys. Rev. Lett.* **46** (1981) 738.

[20] H. Takayama, Y. R. Lin-Liu and K. Maki, *Phys. Rev.* **B21** (1980) 2388.

[21] R. Jackiw and C. Rebbi, *Phys. Rev.* **D13** (1976) 3398.

[22] P. DeVecchia and S. Ferrara, *Nucl. Phys.* **B130** (1977) 93.

[23] N. S. Baaklini, *J. Phys.* **A11** (1978) 2083.

[24] P. Rossi, *Phys. Lett.* **71B** (1977) 45.

[25] G. Semenoff, H. Matsumoto and H. Umezawa, *Phys. Rev.* **D25** (1982) 1054.

[26] G. Oberlechner, M. Umezawa and Ch. Zenses, *Lett. Nuovo Cimento* **23** (1978) 641.

[27] N. J. Papastamatiou, H. Matsumoto and H. Umezawa, *Phys. Rev.* **D29** (1984) 2979.

[28] Y. Leblanc, H. Matsumoto, H. Umezawa and F. Mancini, *Phys. Rev.* **B30** (1984) 5958.

[29] A. R. Bishop, D. K. Campbell, P. S. Lomdahl, B. Horowitz and S. R. Phillpot, *Phys. Rev. Lett.* **52** (1984) 671.

[30] M. Salerno and A. C. Scott, *Phys. Rev.* **B26** (1982) 2474.

[31] M. Salerno, M. P. Soerensen, O. Skovgaard and P. L. Christiansen, *Wave Motion* **5** (1983) 49.

[32] U. Krause, J. P. Kuska and R. Wedell, Monovacancy formation energies in cubic crystals, *Preprint*, 1989.

[33] G. tHooft, *Nucl. Phys.* **B79** (1974) 276.

[34] G. Papini, In B. Kursunoglu, S. L. Mintz and A. Perlmutter, editors, *High Energy Physics*, (Plenum, New York, 1985).

[35] J. P. Elliot, *Proc. Roy. Soc.* **A245** (1958) 128.

[36] A. Arima and F. Iachello, *The Interacting boson Model*, (Cambridge Press, Cambridge, 1987).

[37] M. Mukerjee and Y. Nambu, *Ann. Phys.* **191** (1989) 143.

[38] X. Q. Zhu, F. C. Khanna and H. Umezawa, *Phys. Rev.* **C43** (1991) 2891.

[39] H. Fröhlich, *Intern. J. Quantum Chem.* **2** (1968) 641.

[40] A. S. Davidov, *Phys. Scripta* **20** (1979) 387.

[41] E. Del Giudice, G. Preparata and G. Vitiello, *Phys. Rev. Lett.* **61** (1988) 1085.

[42] E. Del Giudice, S. Doglia, M. Milani and G. Vitiello, *Nucl. Phys.* **B251** (1985) 375.

[43] K. H. Pribram, *Brain and Perception*, (Lawrence Erlbaum Associates, New Jersey, 1991).

[44] J. E. Burns, *The Journal of Mind and Behavior* **11** (1990) 153.

[45] L. M. Ricciardi and H. Umezawa, *Kybernetik* **4** (1967) 44.

[46] C. I. J. M. Stuart, Y. Takahashi and H. Umezawa, *J. Theor. Biol.* **71** (1978) 605.

[47] M. Jibu and K. Yasue, *"A Physical Picture of Umezawa's quantum brain dynamics"*, in Cybernetics and Systems Research 1992, ed. R. Trappl, World Science Publisher, p. 797.

7

Thermo Field Dynamics

7.1 Introduction

We have shown in previous chapters some magical power of quantum field theory in describing a huge variety of the forms that nature manifests. This power originates from the infinite number of degrees of freedom in a quantum field system which gives rise to an infinite number of choices of inequivalent state vector spaces. Each different choice for these spaces appears as a different choice of vacuum, which differ from each other through the form of the particle condensation in the vacuum. The properties which manifest through particle condensation in the vacuum are said to be of macroscopic nature, while those of excited states are called the states of microscopic quasi-particles. In this way a system of quantum fields carries both microscopic and macroscopic properties. Furthermore, we learned that macroscopic objects created by particle condensation carry also quantum mechanical degrees of freedom. Thus, a quantum field system can exhibit the quantum field particles, quantum mechanical and classical degrees of freedom.

When the form of the condensation of particles in a vacuum violates a symmetry of the Hamiltonian, the symmetry is said to be spontaneously broken. Through this occurrence many kinds of macroscopic order emerge. Then, the original invariant property of the Hamiltonian creates a set of continuously degenerate eigenstates of the Hamiltonian, leading to the appearance of degenerate vacua. Presence of these degenerate vacua give rise to energy gapless quantum modes called the NG modes.

Although this power of quantum field theory in describing the rich choices of phase of a system is a welcome aspect of the theory, this flexibility is also a weak point of the theory in the sense that it does not provide a unique answer. In reality each phase of a system is chosen by particular choice of thermodynamical parameters such as temperature, pressure, etc. It is therefore a natural question to ask if some of the degrees of freedom associated with vacuum in quantum field theory could act as a thermal degree of freedom. It is almost obvious that the vacuum in the usual field theory does not carry a thermal degree of freedom, because the usual theory describes the dynamical properties only; the Hamiltonian H in the usual quantum field theory gives rise to the dynamical energy, but it does not give the heat energy. It is apparent that we need some thermal degree of freedom to amend this situation. We now try to extend quantum field theory by adding the thermal degree of freedom to the vacuum.

Our starting point is the consideration presented in section 2.5 which suggests that we take advantage of the two features of the quantum field theoretical formalism for the two mode squeezed states. First, as was explained in section 2.5, the TFD mechanism induces thermal-like noise in two mode squeezed states created by the thermal Bogoliubov transformation. This was extended to quantum field theory in section 3.2. In this way the TFD mechanism and its Bogoliubov transformation create thermal noise in pure states. This feature of the TFD approach is useful for our purposes. The second feature to be used in our attempt is the equivalence between pure state expectation values in the vacuum state created by this TFD mechanism and the thermal average in statistical mechanics for mixed states. This was shown in the relation (2.96) of section 2.4. The vacuum expectation value of operators of one part (that is, the nontilde part) in the system of the two mode squeezed state is equal to the thermal (or ensemble) average in statistical mechanics. This provides a proof for the fact that the thermal quantum theory to be introduced in this chapter coincides with statistical mechanics. This quantum field theory is called Thermo Field Dynamics (TFD).

Since this is a new development, it remains to be seen how much of statistical mechanics can be inherited by TFD. It has frequently been seen in the history of physics that a new formalism built on a broad subject develops a new view upon which new physics is built. It is therefore worthwhile pursuing this pure state view for thermal physics.

Furthermore, there are some advantages to this quantum field formalism that are immediately expected. Statistical mechanics is usually formulated for a system of finite volume and the infinite volume is considered in the thermodynamical limit. On the other hand, as was explained in subsection 6.5.1, quantum field theory deals with a system of infinite volume and objects with a finite volume are treated as ones enclosed by self-consistently created boundary surfaces.

Another possible merit is concerned with the fact that the presence of an infinite number of degrees of freedom may lead us to a thermal theory for an isolated system. In traditional thermodynamics, thermal control is usually performed by a heat bath which is attached to the system under consideration. Since the heat bath should not be disturbed by a feedback effect from the system, it is required to have an infinite number of degrees of freedom. With infinite degrees of freedom quantum field theory is able to include the heat bath, treating the entire system as an isolated system. In this consideration we borrow the terminology from thermodynamics, in which systems are classified into isolated, closed and open ones. An isolated system has no communication with the outside at all. A closed system admits only energy (including heat) flow from the outside of the system, while an open system allows both energy and particle flow. In other words, an open system is coupled to heat bath and particle reservoir, but only heat bath is attached to a closed system.

A quantum field formalism for isolated systems, suggested in the preceding paragraph, may have even some feature of closed or open systems,
because its degrees of freedom are inexhaustible. Any observation in quantum field theory covers a finite domain which is a part of an isolated system.
The outside domain carries an infinite number of degrees of freedom and
may acts as a heat bath or particle reservoir. We will find in a later chapter that, even in a thermal equilibrium state, excited states (that is, the
quasi-particle states) are dissipative, though the vacuum state is stable.
This feature of TFD is in sharp contrast to the usual quantum field theory
without the thermal degree of freedom; in the latter, not only the vacuum,
but also the quasi-particles are stable.

The aspect of TFD applicable to an isolated system of quantum fields is
particularly valuable for several physical reasons. The first is in the physics
of evolution of the Universe. It is commonly assumed that this evolution
has been carried by both dynamical and thermal effects. According to modern physics the fundamental entities in nature are quantum fields such as
quarks, leptons, gauge fields and gravitational fields. These quantum fields
are supposed to be the ones which control the dynamical development.
Since the thermal effects are also involved in the evolution, we need a
quantum field theory with the thermal degrees of freedom. A significant
aspect in this thermal problem is the fact that the Universe is an isolated
system, indicating the need for a thermal quantum field theory for isolated
systems.

The second subject is nuclear physics in which the concept of temperature
has been commonly made use of. It is obvious that nuclear reactions are
phenomena in an isolated system.

Once we accept the assumption that thermal effects are involved in some
areas of the physics of isolated systems (such as the Universe or the nuclear
reactions), we cannot avoid the view that the thermal degree of freedom is
an intrinsic part of quantum field theory. We pursue this view in this book.

Then it appears inevitable to expect that this thermal quantum field
theory should be applied to other phenomena taking place in an isolated
system. This leads many high energy physicists to the speculation that
there may appear for a short time high temperature intermediate states in
high energy particle reactions which occur in an isolated system. In other
words, the high energy particle reactions provide us with a way to answer
to the question as to whether or not thermal effects show up even in the
phenomena of isolated systems in such a fundamental level as the quark
physics. To check this point we need some theoretical results to be compared with experiment. This motivates us to attempt to elaborate thermal
quantum field theories. There have been many efforts made by physicists
to study thermal effects in high energy particle reactions. Most calculations
so far have been confined to equilibrium calculations of self-energy and free
energy with the hope that the thermal intermediate state can be treated
approximately as an equilibrium state. However, any study of particle reac-

tions ultimately requires the calculation of transition matrices. Therefore, we need a theory with which one can calculate transition matrices including thermal effects. Such a theory has yet to be established.

It can be argued that the application of statistical mechanics to high energy particle reactions is motivated, not by thermal effects, but by the need to perform an average over many uncontrollable parameters. This view does not really make much difference to the above fundamental view for thermal effects, because TFD is based on the relation (2.96) in section 2.4 which implies the equivalence between the vacuum expectation value of operators of one part (that is, the nontilde part) in two mode systems and the thermal (or ensemble) average in statistical mechanics. The fact is that both views suggest the need to study thermal effects in phenomena controlled by fundamental quantum fields.

To avoid a possible misunderstanding a note may be in order. Although TFD is expected to be applicable to phenomena in an isolated system, it should be useful to closed or open systems, too. TFD for equilibrium situations is an example of closed systems.

7.2   Hermitian TFD for Free Fields

7.2.1   THERMAL DOUBLETS AND HAMILTONIAN

Considering the relation (2.96) in section 2.4 which replaces the ensemble average by the vacuum expectation value, TFD starts with the formulation for the two mode squeezed state presented in section 3.2. Let us now summarize it for free fields.

Consider a generator consisting of two commuting sets of oscillators:

$$[a_k, a_l^\dagger] = [\tilde{a}_k, \tilde{a}_l^\dagger] = \delta(\vec{k} - \vec{l}), \tag{7.1}$$

$$[a_k, \tilde{a}_l] = [a_k, \tilde{a}_l^\dagger] = 0. \tag{7.2}$$

Unless a system is at zero temperature, a_k and $\tilde{a}_k$ are not the annihilation operators, because they do not annihilate ket-vacuum:

$$a_k|0(\theta)\rangle \neq 0, \quad \tilde{a}_k|0(\theta)\rangle \neq 0, \tag{7.3}$$

$$\langle 0(\theta)|a_k^\dagger \neq 0, \quad \langle 0(\theta)|\tilde{a}_k^\dagger \neq 0, \tag{7.4}$$

where $|0(\theta)\rangle$ and $\langle 0(\theta)|$ are the vacua under consideration. Since the choice of annihilation operators is specified by the choice of vacuum whose label is the parameter θ, the annihilation operators are denoted by $\xi_k(\theta)$ and $\tilde{\xi}_k(\theta)$:

$$\xi_k(\theta)|0(\theta)\rangle = \tilde{\xi}_k(\theta)|0(\theta)\rangle = 0, \tag{7.5}$$

$$\langle 0(\theta)|\xi_k^\dagger(\theta) = \langle 0(\theta)|\tilde{\xi}_k^\dagger(\theta) = 0. \tag{7.6}$$

According to the consideration in section 3.2, these operators are related to the original operators $(a_k, \tilde{a}_k)$ through the thermal Bogoliubov transformation:

$$\xi_k(\theta) = c_k a_k - d_k \tilde{a}_k^\dagger, \tag{7.7}$$

$$\tilde{\xi}_k(\theta) = c_k \tilde{a}_k - d_k a_k^\dagger. \tag{7.8}$$

with $c_k = \cosh \theta_k$ and $d_k = \sinh \theta_k$. The number parameters n_k are defined by

$$n_k \delta(\vec{k} - \vec{l}) \equiv \langle 0(\theta)| a_k^\dagger a_l |0(\theta)\rangle, \tag{7.9}$$

which gives $n_k[\delta(\vec{q})]_{\vec{q}=0} = \langle 0(\theta)| a_k^\dagger a_k |0(\theta)\rangle$ $(\vec{q} = \vec{k} - \vec{l})$. Since $(2\pi)^3 [\delta(\vec{q})]_{\vec{q}=0}$ is the spatial integration of $\exp[i\vec{q} \cdot \vec{x}]$ with $\vec{q} = 0$, it is the volume. Thus, $(2\pi)^{-3} n_k$ is the number density. Using the terminology adapted in section 3.1.3, we call n_k the number density.

As we did in section 3.2, the relations (7.7) and (7.8) can be put in the form

$$\begin{bmatrix} a_k \\ \tilde{a}_k^\dagger \end{bmatrix}^\mu = B_k^{-1}(\theta)^{\mu\nu} \begin{bmatrix} \xi(\theta) \\ \tilde{\xi}_k^\dagger(\theta) \end{bmatrix}^\nu, \tag{7.10}$$

$$[a_k^\dagger, -\tilde{a}_k]^\mu = [\xi_k^\dagger(\theta), -\tilde{\xi}_k(\theta)]^\nu B(\theta)^{\nu\mu} \tag{7.11}$$

with the Bogoliubov matrix $B(\theta)$ of the form

$$B(\theta) = \begin{bmatrix} c & -d \\ -d & c \end{bmatrix}. \tag{7.12}$$

This motivates us to introduce the thermal doublet notation:

$$a_k^1 = a_k, \ a_k^2 = \tilde{a}_k^\dagger, \tag{7.13}$$

$$\bar{a}_k^1 = a_k^\dagger, \ \bar{a}_k^2 = -\tilde{a}_k, \tag{7.14}$$

and

$$\xi_k^1 = \xi_k, \ \xi_k^2 = \tilde{\xi}_k^\dagger, \tag{7.15}$$

$$\bar{\xi}_k^1 = \xi_k^\dagger, \ \bar{\xi}_k^2 = -\tilde{\xi}_k, \tag{7.16}$$

Use of this notation allows us to rewrite (7.10) and (7.11) as

$$a_k^\mu = B_k(\theta)^{\mu\nu} \xi_k(\theta)^\nu, \tag{7.17}$$

$$\bar{a}_k^\mu = \bar{\xi}_k(\theta)^\nu B_k^{-1}(\theta)^{\nu\mu}. \tag{7.18}$$

In the following the thermal doublet notation will be extensively used.

According to section 3.2 this transformation is generated as

$$\xi_k(\theta) = U_B(\theta) a_k U_B^{-1}(\theta), \tag{7.19}$$

$$\tilde{\xi}_k(\theta) = U_B(\theta) \tilde{a}_k U_B^{-1}(\theta), \tag{7.20}$$

where

$$U_B(\theta) = \exp[iG_B(\theta)] \tag{7.21}$$

with

$$G_B(\theta) = i \int d^3k \, \theta_k [a_k \tilde{a}_k - \tilde{a}_k^\dagger a_k^\dagger]. \tag{7.22}$$

Using this generator it was shown in section 3.2 that a $a_k \tilde{a}_k$-pair (called the thermal pair) condensate is present in the vacuum $|0(\theta)\rangle$. There the Fock space built on the vacuum $|0(\theta)\rangle$ was denoted by $\mathcal{H}(\theta)$. Then, it was shown that this thermal pair condensate creates the parameterized Fock spaces $\mathcal{H}(\theta)$ which form an inequivalent set, meaning that two spaces $\mathcal{H}(\theta)$ and $\mathcal{H}(\theta')$ are inequivalent to each other when $\theta \neq \theta'$. When we apply this formulation to thermal physics, θ is the parameter classifying thermal states; for example, two different temperatures correspond to two different choices of θ.

As it was discussed in section 3.3, the generator G_B is an anomalous operator and therefore its definition needs particular care, such as using the smearing trick, though its use frequently helps our intuitive understanding of the structure of the vacuum. We emphasize that the relations (7.7) and (7.8) are well defined; the operations of a_k and $\tilde{a}_k$ on vectors in $\mathcal{H}(\theta)$ are defined through these relations. Since a_k and $\tilde{a}_k$ do not annihilate the vacuum $|0(\theta)\rangle$, they should not be called the annihilation operators.

Since $U_B(\theta)$ is a unitary operator, $\xi_k^\dagger$ and $\tilde{\xi}_k^\dagger$ are respectively the Hermitian conjugate of ξ_k and $\tilde{\xi}_k$, because $a_k^\dagger$ and $\tilde{a}_k^\dagger$ are Hermitian conjugate to a_k and $\tilde{a}_k$, respectively. Therefore, in this formulation the dagger symbol means the Hermitian conjugation. Although this comment might seem to be obvious, it is of significance because later we generalize the definition of dagger conjugation. In subsection 2.4.3, we pointed out that there are two parameters which can make the bra-vacuum not the Hermitian conjugate of the ket-vacuum. They were α and (c_1/c_2). The parameters (α, c_1, c_2) appeared in the definition of vacua in (2.90) and (2.91). The representation in which the bra-vacuum is the Hermitian conjugate of the ket-vacuum is called the unitary representation and is given by the choice $\alpha = 1/2$ and $c_1/c_2 = 1$. TFD in which U_B is unitary is this unitary representation. In the unitary representation dagger conjugation is Hermitian conjugation. TFD in the unitary representation is called the unitary TFD.

The unitary representation in TFD does not mean that the Bogoliubov matrix is unitary. According to the definition of thermal doublets in (7.16), the second relation in the Bogoliubov transformation, that is the relation in (7.18), reads as

$$[\xi(\theta)^\mu]^\dagger = [a^\nu]^\dagger [\tau_3 B^{-1}(\theta)\tau_3]^{\nu\mu}. \tag{7.23}$$

Here use was made of the Pauli matrices:

$$\tau_1 = \begin{bmatrix} 0 & 1 \\ 1 & 0 \end{bmatrix}, \quad \tau_2 = \begin{bmatrix} 0 & -i \\ i & 0 \end{bmatrix}, \quad \tau_3 = \begin{bmatrix} 1 & 0 \\ 0 & -1 \end{bmatrix}. \tag{7.24}$$

On the other hand Hermitian conjugate of the first relation of the Bogoliubov transformation, that is (7.17), gives

$$[\xi(\theta)^\mu]^\dagger = [a^\nu]^\dagger [B^\dagger(\theta)]^{\nu\mu}. \tag{7.25}$$

Comparing this with (7.23), we find that *the condition on the Bogoliubov matrix for a unitary representation is*

$$B^\dagger(\theta) = \tau_3 B^{-1}(\theta)\tau_3. \tag{7.26}$$

It was shown in subsection 2.3.2 that the time independence of the thermal vacuum requires a particular relation between the energy of nontilde and tilde systems. The argument goes as follows. Suppose that the free Hamiltonian of the system under consideration is

$$H_{tot} = \int d^3k[\omega_k^1 a_k^\dagger a_k + \omega_k^2 \tilde{a}_k^\dagger \tilde{a}_k]. \tag{7.27}$$

Then the time dependence of these oscillator operators is $a_k \exp[-i\omega_k^1 t]$ and $\tilde{a}_k \exp[-i\omega_k^2 t]$. Therefore each thermal pair carries the wave function $\exp[i(\omega_k^1 + \omega_k^2)t]$, which depends on time unless $\omega_k^2 = -\omega_k^1$. This shows that the thermal vacuum $|0(\theta)\rangle$ is independent of time t when and only when $\omega_k^2 = -\omega_k^1$. This can also be seen from the fact that the Hamiltonian H_{tot} and the generator G_B do not commute with each other unless $\omega_k^2 = -\omega_k^1$.

When we require that the vacuum be independent of time, we have the following total Hamiltonian for free fields:

$$\hat{H}_0 = \int d^3k\omega_k[a_k^\dagger a_k - \tilde{a}_k^\dagger \tilde{a}_k]. \tag{7.28}$$

This is the free field Hamiltonian in TFD.

The relation $\omega_k^2 = -\omega_k^1$ suggests a picture in which tilde quanta are holes of nontilde quanta in the vacuum. This can be further supported by the following consideration. Suppose a system is in a thermal equilibrium state with temperature T. In order to find how many degrees of freedom are associated with this system, we study how this system responds to an external stimulus. When $T \neq 0$, a certain number of quantum particles are condensed in this system. Therefore, absorption of the external energy by the system occurs in two ways. One is the absorption by excitation of additional quanta. Another is the process in which the external stimulus excites a quantum present in the vacuum condensate, leaving the system with a hole in the condensate. Thus, an external stimulus creates either a quantum or a hole. This explains why we have doubled degrees of freedom.

As was shown in section 3.4.2, the argument in this subsection can be extended to the thermal vacua of fermion fields. According to (3.44) and (3.45) the Bogoliubov transformation for a fermion has the form

$$\begin{aligned} a_k^\mu &= B_k(\theta)^{\mu\nu}\xi_k(\theta)^\nu, &\tag{7.29}\\ \bar{a}_k^\mu &= \bar{\xi}_k(\theta)^\nu B_k^{-1}(\theta)^{\nu\mu}, &\tag{7.30} \end{aligned}$$

where the thermal doublet notation is

$$a_k^1 = a_k, \qquad a_k^2 = \tilde{a}_k^\dagger, \tag{7.31}$$

$$\bar{a}_k^1 = a_k^\dagger, \qquad \bar{a}_k^2 = \tilde{a}_k, \tag{7.32}$$

and

$$\xi_k^1 = \xi_k, \qquad \xi_k^2 = \tilde{\xi}_k^\dagger, \tag{7.33}$$

$$\bar{\xi}_k^1 = \xi_k^\dagger, \qquad \bar{\xi}_k^2 = \tilde{\xi}_k. \tag{7.34}$$

Note the positive sign in definition of $\bar{a}_k^2$ and $\bar{\xi}_k^2$, which is contrary to the minus sign for bosonic case.

The Bogoliubov matrix is given by (3.46):

$$B_k(\theta) = \begin{bmatrix} c_k & -d_k \\ d_k & c_k \end{bmatrix}. \tag{7.35}$$

with $c_k = \cos\theta_k$ and $d_k = \sin\theta_k$. The number density parameter is given by $n_k[\delta(\vec{q})]_{\vec{q}=0} \equiv \langle 0|a_k^\dagger a_k|0\rangle$. We have $n_k = d_k^2$. Note that this Bogoliubov matrix does satisfy the condition for the Hermitian representation (7.26).

This transformations in (7.29) and (7.30) is induced by $U_B(\theta)$ as

$$\xi_k(\theta) = U_B(\theta) a_k U_B^{-1}(\theta), \tag{7.36}$$

$$\tilde{\xi}_k(\theta) = U_B(\theta) \tilde{a}_k U_B^{-1}(\theta), \tag{7.37}$$

where $U_B(\theta)$ was given in (3.50):

$$U_B(\theta) = \exp\left[-\int d^3k\, \theta_k[a_k\tilde{a}_k - \tilde{a}_k^\dagger a_k^\dagger]\right]. \tag{7.38}$$

7.2.2   THE TILDE CONJUGATION RULES

In TFD every dynamical degree of freedom is doubled; to any operator A is associated its tilde conjugate $\tilde{A}$. Therefore we need the rule of tilde conjugation which was briefly touched in section 2.3. Here we set up the structure of the tilde conjugation in more detail by extracting from the above consideration some basic properties of TFD.

First note that the wave function of the a-particle behaves as $\exp[-i\omega_k t]$, while the one of $\tilde{a}$-particle is $\exp[i\omega_k t]$ according to the above Hamiltonian. Therefore, the wave function of $\tilde{a}_k^\dagger$ is $\exp[-i\omega_k t]$, which is the same as the wave function of a_k. This is very significant, because, in order for the Bogoliubov transformation to be able to mix a_k with $\tilde{a}_k^\dagger$ at any time, the wave function of a and that of $\tilde{a}^\dagger$ should have the same form.

Comparing the wave function of a_k with that of $\tilde{a}_k$, we find that the i changes its sign, and therefore that a c-number becomes its complex conjugate under the tilde conjugation.

We now recall that the number parameter is given by the vacuum expectation value of the operator $a_k^\dagger a_k$ which is equal to the vacuum expectation value of $\tilde{a}_k^\dagger \tilde{a}_k$. Thus we consider $\tilde{a}_k^\dagger \tilde{a}_k$ as the tilde conjugate of $a_k^\dagger a_k$. The ordering in this operator is important because a_k and $a_k^\dagger$ do not commute with each other. We thus assume that the tilde conjugation does not change the ordering among operators. We recall also that the thermal vacua are states with $a\tilde{a}$-pair condensate. Therefore, they are invariant under the tilde conjugation.

Note that there is a self-consistency condition for the Bogoliubov transformation; the Bogoliubov transformation should be invariant under the tilde conjugation. This leads us to the following condition for double tilde conjugation: $\tilde{\tilde{a}}_k = \sigma a_k$. Here σ is the sign constant defined as

$$\sigma = \begin{cases} 1 & \text{for boson} \\ -1 & \text{for fermion} \end{cases} \tag{7.39}$$

For fermion we can make a different choice of U_B which gives $\tilde{\tilde{a}}_k = a_k$. Since this choice does not influence physical answers in any way, we proceed with the choice already made in this section.

In brief, the basic relations in TFD can now be stated as follows: To any operator A is associated its tilde conjugate $\tilde{A}$ obeying these tilde conjugation rules:

$$(AB)^\sim = \tilde{A}\tilde{B}, \tag{7.40}$$
$$(c_1 A + c_2 B)^\sim = c_1^* \tilde{A} + c_2^* \tilde{B}, \tag{7.41}$$
$$(A^\dagger)^\sim = \tilde{A}^\dagger, \tag{7.42}$$
$$(\tilde{A})^\sim = \sigma A, \tag{7.43}$$
$$|0(\theta)\rangle^\sim = |0(\theta)\rangle, \tag{7.44}$$
$$\langle 0(\theta)|^\sim = \langle 0(\theta)|. \tag{7.45}$$

Here c_1 and c_2 are any two c-numbers and A and B stand for any two operators. It may be useful to recall here the consideration made in subsection 2.4.3, in which it was shown that the rule of replacing a c-number with its complex conjugate together with the invariance of the vacua under tilde conjugation eliminates the trivial phase freedom in structure of the bra- and ket-vacuum in (2.90) and (2.91).

As we will see in the next section, these tilde conjugation rules form a basic part of the TFD formalism.

Exercise 1 Examine the invariance of the Bogoliubov transformation under the tilde conjugation.

With use of the tilde conjugation rules we can put the free Hamiltonian (7.28) in the following form:

$$\hat{H}_0 = H_0 - \tilde{H}_0. \tag{7.46}$$

with

$$H_0 = \int d^3k \, \omega_k a_k^\dagger a_k. \tag{7.47}$$

In TFD all the dynamical observables are made of nontilde a-operators only. The expectation value of H_0 is the dynamical energy. However, since the Hamiltonian should generate the time change both in nontilde and tilde operators, the full Hamiltonian is not H_0 but $\hat{H}_0$. The eigenvalue of $\hat{H}_0$ is called the hat-energy.

Although the tilde operators do not participate in dynamical observables, thermal observables such as heat energy, entropy and so on need both the nontilde and tilde operators. This will be discussed in later sections and later chapters.

Since the thermal doublet notation is extensively used in TFD, we here summarize them for both boson and fermion:

$$a_k^1 = a_k, \qquad a_k^2 = \tilde{a}_k^\dagger, \tag{7.48}$$

$$\bar{a}_k^1 = a_k^\dagger, \qquad \bar{a}_k^2 = -\sigma \tilde{a}_k, \tag{7.49}$$

and

$$\xi_k^1 = \xi_k, \qquad \xi_k^2 = \tilde{\xi}_k^\dagger, \tag{7.50}$$

$$\bar{\xi}_k^1 = \xi_k^\dagger, \qquad \bar{\xi}_k^2 = -\sigma \tilde{\xi}_k, \tag{7.51}$$

where σ is plus for boson and minus for fermion.

The relations $\bar{a}$ and $\bar{\xi}$ are expressed as

$$\bar{a}^\mu = a^{\nu,\dagger} [\tau_\sigma]^{\nu\mu}, \tag{7.52}$$

$$\bar{\xi}^\mu = \xi^{\nu,\dagger} [\tau_\sigma]^{\nu\mu}, \tag{7.53}$$

where the matrix τ_σ is

$$\tau_\sigma = \begin{bmatrix} 1 & 0 \\ 0 & -\sigma \end{bmatrix}. \tag{7.54}$$

The condition for the unitary representation (7.26) is extended as

$$B^\dagger(\theta) = \tau_\sigma B^{-1}(\theta) \tau_\sigma. \tag{7.55}$$

The general form of the Bogoliubov matrix in the unitary representation is

$$B_k(\theta) = \begin{bmatrix} c_k & -d_k \\ -\sigma d_k & c_k \end{bmatrix}. \tag{7.56}$$

where c_k and d_k are conditioned by the relation

$$c_k^2 - \sigma d_k^2 = 1. \tag{7.57}$$

The definition of the thermal doublets are extended to any operator A which consists of a_k and $a_k^\dagger$ only:

$$A^1 = A, \quad A^2 = \tilde{A}^\dagger, \tag{7.58}$$
$$\bar{A}^1 = A^\dagger, \quad \bar{A}^2 = -\sigma\tilde{A}, \tag{7.59}$$

where σ is plus for bosonic A and minus for fermionic A. Here bosonic (fermionic) A means a sum of product terms which contain even (odd) number of fermion fields and any number of boson fields.

7.2.3   NON HERMITIAN REPRESENTATION OF TFD

Up to now we considered a particular form of the Bogoliubov matrix. However, any transformation which mixes a_k with $\tilde{a}_k^\dagger$ and whose generator commutes with the Hamiltonian, can be used for the creation of thermal vacua. Therefore, it is interesting to ask for the widest choice of the generator. Since we are planning to make use of perturbation calculations, we are interested in unperturbed free fields. In this section we study the widest choice of the Bogoliubov transformation for free fields. This study is useful in the next section when we formulate a perturbative calculation for quantum fields with interactions.

Our starting point is the free Hamiltonian in (7.46):

$$\hat{H}_0 = H_0 - \tilde{H}_0 \tag{7.60}$$
$$= \int d^3k\,\omega_k[a_k^\dagger a_k - \tilde{a}_k^\dagger \tilde{a}_k]. \tag{7.61}$$

Let us denote a general choice of the Bogoliubov generator by $\hat{G}$:

$$\xi_k(\theta) = e^{i\theta\hat{G}}a_k e^{-i\theta\hat{G}}, \qquad \xi_k^\dagger(\theta) = e^{i\theta\hat{G}}a_k^\dagger e^{-i\theta\hat{G}},$$
$$\tilde{\xi}_k(\theta) = e^{i\theta\hat{G}}\tilde{a}_k e^{-i\theta\hat{G}}, \qquad \tilde{\xi}_k^\dagger(\theta) = e^{i\theta\hat{G}}\tilde{a}_k^\dagger e^{-i\theta\hat{G}}. \tag{7.62}$$

Here is a significant note. Although the dagger symbol attached to a_k and $\tilde{a}_k$ means the Hermitian conjugate, the dagger symbol attached to other operators does not necessarily means the Hermitian conjugate: for example, $\xi^\dagger$ and $\tilde{\xi}^\dagger$ are defined through the above relations. In general the dagger conjugate of an operator A is defined through expression of A in terms of a_k and $\tilde{a}_k$.

There are three requirements for $\hat{G}$. The first is that it induces the transformation of the form

$$\xi_k(\theta)^\mu = B_k(\theta)^{\mu\nu}a_k^\nu, \tag{7.63}$$
$$\bar{\xi}_k(\theta)^\mu = \bar{a}_k^\nu B_k^{-1}(\theta)^{\nu\mu}, \tag{7.64}$$

where the definition of the thermal doublets is same as the one given in the last subsection.

The second is that *it does commute with the $\hat{H}_0$*:

$$[\hat{H}_0, \hat{G}] = 0. \tag{7.65}$$

The third is the requirement that the thermal vacua are invariant under the tilde conjugation. The invariance of the vacua under the tilde conjugation requires that $\hat{G}$ changes its sign under the tilde conjugation:

$$(\hat{G})^{\sim} = -\hat{G}. \tag{7.66}$$

These three conditions are used in identifying the $\hat{G}$-symmetry.

Since $\hat{G}$ should generate a linear mixing, it should be bilinear in oscillator operators. The only bilinear products which commute with $\hat{H}_0$ are the following four terms: $a_k\tilde{a}_k, \tilde{a}_k^{\dagger}a_k^{\dagger}, a_k^{\dagger}a_k, \tilde{a}_k^{\dagger}\tilde{a}_k$. Therefore, $\theta\hat{G}$ is of the following form:

$$\theta\hat{G} = \int d^3k \sum_{i=1}^{4} \theta_{i,k}\hat{G}_{i,k} \tag{7.67}$$

with

$$\hat{G}_{1,k} = i(a_k\tilde{a}_k - a_k^{\dagger}\tilde{a}_k^{\dagger}) \tag{7.68}$$

$$\hat{G}_{2,k} = i(a_k\tilde{a}_k + a_k^{\dagger}\tilde{a}_k^{\dagger}) \tag{7.69}$$

$$\hat{G}_{3,k} = i(a_k^{\dagger}a_k + \tilde{a}_k^{\dagger}\tilde{a}_k) \tag{7.70}$$

$$\hat{G}_{4,k} = a_k^{\dagger}a_k - \tilde{a}_k^{\dagger}\tilde{a}_k. \tag{7.71}$$

However, the operator $\hat{G}_{4,k}$, proportional to $\hat{H}_0(\theta)$, commutes with the other $\hat{G}_{i,k}$ ($i = 1 \sim 3$) and also does not change the thermal vacuum. We therefore do not need to consider it. Then, we have three kinds of parameters $\theta_{i,k}, i = 1, 2, 3$. The generators $\hat{G}_{i,k}$ ($i = 1 \sim 3$) satisfy the SU(1,1) algebra.

Note that $\hat{G}_{2,k}$ and $\hat{G}_{3,k}$ are anti Hermitian, a result of the requirement (7.66).

In subsection 2.4.3 it was shown that there are three parameters, $\alpha, c_1/c_2, f$ in the definition of thermal vacua when we do not require the bra-vacuum to be the Hermitian conjugate of the ket-vacuum. This explains why we have three kinds of parameters $\theta_{i,k}; i = 1, 2, 3$.

It is a tedious task to calculate the matrix $B_k(\theta)$ by means of the relations in (7.62). To calculate this matrix for a boson field, it is useful to rewrite $\exp[i\theta\hat{G}]$ as

$$\exp[i\theta\hat{G}] = \exp[i\chi_1\hat{G}_3]\exp[-i\chi_2\hat{G}_2]\exp[i\chi_3\hat{G}_3]\exp[i\chi_2\hat{G}_2]\exp[-i\chi_1\hat{G}_3] \tag{7.72}$$

with the short-handed notation

$$\chi_i\hat{G}_j = \int d^3k \, \chi_{i,k}\hat{G}_{j,k} \tag{7.73}$$

and

$$\tanh \chi_{1,k} = \frac{\theta_{2,k}}{\theta_{1,k}} \tag{7.74}$$

$$\tan 2\chi_{2,k} = \frac{\sqrt{\theta_{1,k}^2 - \theta_{2,k}^2}}{\theta_{3,k}} \tag{7.75}$$

$$\chi_{3,k}^2 = \theta_{3,k}^2 + \theta_{1,k}^2 - \theta_{2,k}^2. \tag{7.76}$$

We then obtain

$$B_k(\theta)^{\mu\nu} =$$
$$\begin{bmatrix} \cosh \chi_{3,k} + \sinh \chi_{3,k} \cos 2\chi_{2,k} & -e^{-2\chi_{1,k}} \sinh \chi_{3,k} \sin 2\chi_{2,k} \\ -e^{2\chi_{1,k}} \sinh \chi_{3,k} \sin 2\chi_{2,k} & \cosh \chi_{3,k} - \sinh \chi_{3,k} \cos 2\chi_{2,k} \end{bmatrix}. \tag{7.77}$$

This form of the matrix $B_k(\theta)$ is to be compared with that commonly used in the literature on TFD (see for example [1]) where a different parameterization is used, namely,

$$B_k(\theta)^{\mu\nu} = (1 + n_k)^{1/2} e^{s_k \tau_3} \begin{bmatrix} 1 & -f_k^{\alpha_k} \\ -f_k^{1-\alpha_k} & 1 \end{bmatrix}, \tag{7.78}$$

where $f_k = n_k/[1 + n_k]$ which gives $n_k = f_k/[1 - f_k]$. This matrix contains the three parameters, n_k, α_k and s_k. Here τ_i ($i = 1 \sim 3$) are the Pauli matrices. The parameter n_k is the number density defined by

$$n_k(\theta)\delta(\vec{k} - \vec{l}) = \langle 0(\theta)|a_k^\dagger a_l|0(\theta)\rangle. \tag{7.79}$$

The parameter α_k is the parameter α mentioned in (2.90) and (2.91). This freedom corresponds to the one appearing in the density matrix formalism due to the cyclic property of ρ under the trace formula when we write

$$\rho = \rho^{1-\alpha}\rho^\alpha. \tag{7.80}$$

Thus we usually choose α to be independent of $\vec{k}$ and denote it by α.

Comparing (7.77) with (7.78), we find the following one-to-one correspondence between $\{n_k, \alpha_k, s_k\}$ and θ_i (or χ_i):

$$n_k = \sinh^2 \chi_{3,k} \sin^2 2\chi_{2,k} \tag{7.81}$$

$$\alpha_k = \frac{-2\chi_{1,k} + \frac{1}{2}\ln(\sinh^2 \chi_{3,k} \sin^2 2\chi_{2,k})}{\ln\left(\frac{\sinh^2 \chi_{3,k} \sin^2 2\chi_{2,k}}{1+\sinh^2 \chi_{3,k} \sin^2 2\chi_{2,k}}\right)} \tag{7.82}$$

$$s_k = \frac{1}{2}\ln\left(\frac{\cosh \chi_{3,k} + \sinh \chi_{3,k} \cos 2\chi_{2,k}}{\cosh \chi_{3,k} - \sinh \chi_{3,k} \cos 2\chi_{2,k}}\right). \tag{7.83}$$

The above consideration can be easily extended to fermions. We present the general structure of the Bogoliubov transformation both for bosons and fermions. The transformations are the same as those in (7.63) and (7.64):

$$\xi_k(\theta)^\mu = B_k(\theta)^{\mu\nu} a_k^\nu, \tag{7.84}$$

$$\bar\xi_k(\theta)^\mu = \bar a_k^\nu B_k^{-1}(\theta)^{\nu\mu}, \tag{7.85}$$

where the definition of the thermal doublets is the same as the one given in the last subsection.

The Bogoliubov matrix is

$$B_k(\theta)^{\mu\nu} = (1 + \sigma n_k)^{1/2} e^{s_k \tau_3} \begin{bmatrix} 1 & -f_k^\alpha \\ -\sigma f_k^{1-\alpha} & 1 \end{bmatrix}, \tag{7.86}$$

where σ is plus for boson and minus for fermion. The f_k is

$$f_k = \frac{n_k}{1 + \sigma n_k}. \tag{7.87}$$

which gives $n_k = f_k/[1 - \sigma f_k]$. We chose α_k to be independent of $\vec k$ and denote it by α.

The inverse of the above matrix is

$$B_k^{-1}(\theta)^{\mu\nu} = (1 + \sigma n_k)^{1/2} \begin{bmatrix} 1 & f_k^\alpha \\ \sigma f_k^{1-\alpha} & 1 \end{bmatrix} e^{-s_k \tau_3}, \tag{7.88}$$

Exercise 2 When we want to emphasize the α-dependence of B_k, we denote it by $B_{\alpha k}$. Prove then the following relation [2]:

$$B_{\alpha_1 k} = W_{21k}^{-1} B_{\alpha_2 k} W_{21k} \tag{7.89}$$

with

$$W_{21k} = f_k^{\frac{(\alpha_2 - \alpha_1)}{2}\tau_3}. \tag{7.90}$$

With this rule we can use the freedom of changing α if desired.

7.2.4 SPONTANEOUS BREAKDOWN OF $\hat G$-SYMMETRY

Note now that $\hat G$ does not annihilate the vacuum:

$$\theta \hat G |0(\theta)\rangle \neq 0. \tag{7.91}$$

This means that the vacuum is not invariant under the transformation generated by $\hat G$, although the Hamiltonian $\hat H_0$ is invariant under this transformation because $\hat H_0$ commutes with $\hat G$. Therefore the $\hat G$-symmetry is spontaneously broken.

The structure of $\hat G$ indicates that the thermal vacua are states with a $a\tilde a$-pair condensate. A change from $|0(\theta)\rangle$ to $|0(\theta')\rangle$ is made by modifying the distribution of this pair condensate.

As was discussed in the last subsection, all the dynamical observables are made of nontilde a-operators only. Thus the expectation value of H_0 is the dynamical energy. However, the Hamiltonian should generate the time change both in nontilde and tilde operators, so the Hamiltonian is not H_0 but $\hat{H}_0$. The eigenvalue of $\hat{H}_0$ is called the hat-energy. The a-field carries the hat-energy ω_k, while the $\tilde{a}_k$ carries the hat-energy $(-\omega_k)$. Thus the thermal pair $a\tilde{a}$ in the vacuum carries zero hat-energy. This is the NG mode associated with the symmetry breakdown of the $\hat{G}$-symmetry.

We have the Fock space built on each vacuum $|0(\theta)\rangle$. This is called $\mathcal{H}(\theta)$. These Fock spaces form an inequivalent set with continuous parameters represented by θ. Among these parameters, α_k and s_k have only mathematical relevance, but a change of n_k has a physically relevant effect.

7.2.5 THE THERMAL STATE CONDITION

When the Bogoliubov transformations in (7.84) and (7.85) are considered, the relations

$$\xi(\theta)|0(\theta)\rangle = \tilde{\xi}(\theta)|0(\theta)\rangle = 0, \qquad (7.92)$$
$$\langle 0(\theta)|\xi^\dagger(\theta) = \langle 0(\theta)|\tilde{\xi}^\dagger(\theta) = 0 \qquad (7.93)$$

give

$$[a_k - f_k^\alpha \tilde{a}_k^\dagger]|0(\theta)\rangle = 0, \qquad (7.94)$$
$$[\tilde{a}_k - \sigma f_k^\alpha a_k^\dagger]|0(\theta)\rangle = 0, \qquad (7.95)$$
$$\langle 0(\theta)|[a^\dagger - f_k^{1-\alpha}\tilde{a}_k] = 0, \qquad (7.96)$$
$$\langle 0(\theta)|[\tilde{a}_k^\dagger - \sigma f_k^{1-\alpha} a_k] = 0. \qquad (7.97)$$

This is the simplest form of the relations called thermal state conditions. Here we chose the parameter α_k to be independent of $\vec{k}$.

Let us now define c_k by

$$f_k = e^{c_k} \qquad (7.98)$$

and introduce

$$\hat{M}_c = \int d^3 k\, c_k \hat{N}_k, \qquad (7.99)$$

where

$$\hat{N}_k \equiv [a_k^\dagger a_k - \tilde{a}_k^\dagger \tilde{a}_k]. \qquad (7.100)$$

Note that $\hat{N}_k$ commutes with $\hat{G}$. Therefore,

$$\hat{N}_k = [\xi_k^\dagger(\theta)\xi_k(\theta) - \tilde{\xi}_k^\dagger(\theta)\tilde{\xi}_k(\theta)]. \qquad (7.101)$$

It follows that

$$\hat{N}_k|0(\theta)\rangle = \langle 0(\theta)|\hat{N}_k = 0, \qquad (7.102)$$

which gives

$$\hat{M}_c|0(\theta)\rangle = \langle 0(\theta)|\hat{M}_c = 0. \tag{7.103}$$

On the other hand, we have

$$e^{-\hat{M}_c}a_k e^{\hat{M}_c} = e^{c_k}a_k, \tag{7.104}$$

$$e^{-\hat{M}_c}a_k^\dagger e^{\hat{M}_c} = e^{-c_k}a_k^\dagger, \tag{7.105}$$

$$e^{-\hat{M}_c}\tilde{a}_k e^{\hat{M}_c} = e^{-c_k}\tilde{a}_k, \tag{7.106}$$

$$e^{-\hat{M}_c}\tilde{a}_k^\dagger e^{\hat{M}_c} = e^{c_k}\tilde{a}_k^\dagger. \tag{7.107}$$

Now (7.94) becomes

$$[a_k - e^{-\alpha\hat{M}_c}\tilde{a}_k^\dagger]|0(\theta)\rangle = 0. \tag{7.108}$$

Multiplying both sides of (7.94) with $a_l^\dagger$ and making use of (7.95) gives

$$[a_l^\dagger a_k - e^{\alpha(c_k-c_l)}\tilde{a}_k^\dagger\tilde{a}_l]|0(\theta)\rangle = 0, \tag{7.109}$$

which reads as

$$[a_l^\dagger a_k - e^{-\alpha\hat{M}_c}(\tilde{a}_l^\dagger\tilde{a}_k)^\dagger]|0(\theta)\rangle = 0. \tag{7.110}$$

A similar computation gives

$$[A - e^{-\alpha\hat{M}_c}\tilde{A}^\dagger]|0(\theta)\rangle = 0, \tag{7.111}$$

$$\langle 0(\theta)|[A^\dagger - \tilde{A}e^{-(1-\alpha)\hat{M}_c}] = 0. \tag{7.112}$$

Here A stands for any sum of normal products consisting of a_k and $a_k^\dagger$. These thermal state conditions are useful when A stands for a Heisenberg operator whose dynamical map is expressed in terms of free fields.

When we consider an equilibrium situation with temperature $T = 1/\beta$, we have $f_k = \exp[-\beta\omega_k]$ which gives $c_k = -\beta\omega_k$. In this case we have $\hat{M}_c = -\beta\hat{H}_0$.

Since equilibrium states are stationary, their thermal vacua are invariant under the time translation, meaning that the time translational symmetry is not broken. Therefore, we can extend the consideration of dynamical map of Heisenberg operators in subsection 4.6.1 to TFD for equilibrium situations. Thus, we can see that when the dynamical map of Heisenberg fields can be expressed in terms of free fields, the total Hamiltonian $\hat{H}$ is weakly equal to $\hat{H}_0$ as we saw in chapter 4. Then, the above equations become

$$[A - e^{\beta\alpha\hat{H}}\tilde{A}^\dagger]|0(\theta)\rangle = 0, \tag{7.113}$$

$$\langle 0(\theta)|[A - \tilde{A}^\dagger e^{(1-\alpha)\beta\hat{H}}] = 0, \tag{7.114}$$

in equilibrium situation.

However, when even a slight time dependent effect is present, the dynamical map in terms of quasi-particle fields in TFD requires careful treatment. This will be seen in the next chapter.

7.2.6   ONE BODY PROPAGATORS IN TFD

Knowing the thermal Bogoliubov transformation, calculation of the one body propagator for free fields is an easy task. Consider a free scalar field $\varphi(x)$. We assume that the Bogoliubov parameters θ_i are independent of time.

We first consider a field satisfying the field equation

$$[i\frac{\partial}{\partial t} - \omega(-i\vec{\nabla})]\varphi(x) = 0 \qquad (7.115)$$

with quasi-particle energy $\omega_k = \omega(\vec{k})$. A field of this kind is called type 1.

In TFD we introduce the tilde conjugate of φ and apply the thermal doublet notation in (7.59):

$$\varphi^1 = \varphi, \ \varphi^2 = \tilde{\varphi}^\dagger, \qquad (7.116)$$
$$\bar{\varphi}^1 = \varphi^\dagger, \ \bar{\varphi}^2 = -\sigma\tilde{\varphi}, \qquad (7.117)$$

Then, these thermal doublets are expressed in terms of the oscillator operators a_k^μ as

$$\varphi(x)^\mu = \int \frac{d^3k}{(2\pi)^{3/2}} e^{i[\vec{k}\cdot\vec{x}-\omega_k t]} a_k^\mu$$
$$\bar{\varphi}(x)^\mu = \int \frac{d^3k}{(2\pi)^{3/2}} e^{-i[\vec{k}\cdot\vec{x}-\omega_k t]} \bar{a}_k^\mu. \qquad (7.118)$$

Let us begin with zero temperature. Its one body propagator at zero temperature is denoted by

$$\Delta_0(x-y)^{\mu\nu} = \langle\langle 0|T[\varphi(x)^\mu\bar{\varphi}(y)^\nu]|0\rangle\rangle, \qquad (7.119)$$

where the suffix zero means zero temperature.

The Fourier form of this propagator of the thermal doublet at zero temperature is written as

$$\Delta_0(x-y)^{\mu\nu} = \frac{1}{(2\pi)^4} \int d^4k \, e^{-ik(x-y)} \Delta_0(k)^{\mu\nu}, \qquad (7.120)$$

with

$$\Delta_0(k)^{\mu\nu} = [\Delta_0(k, \epsilon_F\tau_3)]^{\mu\nu} \qquad (7.121)$$
$$= \left[\frac{i}{k_0 - \omega_k + i\epsilon_F\tau_3}\right]^{\mu\nu}, \qquad (7.122)$$

where ϵ_F is the Feynman positive infinitesimal. The only matrix structure comes from τ_3.

Exercise 3 Prove (7.122).

Hint1: The components with $\mu \neq \nu$ vanish.

Hint2: The $\Delta_0(x-y)^{11}$ is same as the one in the usual field theory with no thermal degree of freedom:

$$\Delta_0(x-y)^{11} = \langle\langle 0|T[\varphi(x)\varphi^\dagger(y)]|0\rangle\rangle. \tag{7.123}$$

Its Fourier amplitude is written as $\Delta_0(k, \epsilon_F)^{11}$:

$$\Delta_0(x-y)^{11} = \frac{1}{(2\pi)^4} \int d^4k \, e^{-ik(x-y)} \Delta_0(k, \epsilon_F) \tag{7.124}$$

with

$$\Delta_0(k, \epsilon_F)^{11} = \frac{i}{k_0 - \omega_k + i\epsilon_F}. \tag{7.125}$$

The significant point is that the i-factor and ϵ_F are the only terms which are imaginary.

Hint3: For $\mu = \nu = 2$ we have

$$\Delta_0(x-y)^{22} = -\sigma\langle\langle 0|T[\tilde{\varphi}^\dagger(x)\tilde{\varphi}(y)]|0\rangle\rangle \tag{7.126}$$
$$= -\langle\langle 0|T[\tilde{\varphi}(y)\tilde{\varphi}^\dagger(x)]|0\rangle\rangle. \tag{7.127}$$

Except for the minus sign, this is the tilde conjugate of $\Delta_0(y-x)^{11}$, which is the complex conjugate of $\Delta_0(y-x)^{11}$ because the propagator is a c-number. Note the exchange of x and y. Thus, we have

$$\Delta_0(x-y)^{22} = \frac{1}{(2\pi)^4} \int d^4k \, e^{-ik(x-y)} \Delta_0(k, -\epsilon_F). \tag{7.128}$$

We found that $\Delta_0(x-y)^{22}$ differs from $\Delta_0(x-y)^{11}$ only by the sign of ϵ_F.

The propagator with the thermal vacuum $|0(\theta)\rangle$ is related to Δ_0 through the Bogoliubov transformation $B_k(\theta)$. Thus we have

$$\Delta(x-y)^{\mu\nu} \equiv \langle 0(\theta)|T[\varphi(x)^\mu\tilde{\varphi}(y)^\nu]|0(\theta)\rangle \tag{7.129}$$
$$= \frac{1}{(2\pi)^4} \int d^4k \, e^{ik(x-y)} \Delta(k)^{\mu\nu} \tag{7.130}$$

with

$$\Delta(k)^{\mu\nu} = [B_k^{-1}(\theta)] \left(\frac{i}{k_0 - \omega_k + i\epsilon_F\tau_3} \right) B_k(\theta)]^{\mu\nu}. \tag{7.131}$$

This propagator is obtained from the one in the usual quantum field theory with no thermal degree of freedom by the following simple rule; replace ϵ_F by $\epsilon_F\tau_3$ and operate the Bogoliubov transformation.

Now recall the general expression for the Bogoliubov matrix given in (7.78) in subsection 7.2.3. This expression contained the parameters α and

s_k besides the significant parameter n_k. The above result for the free propagator shows that *the parameter s_k does not contribute to the propagator.* As a matter of fact, s_k influences only the normalization factors of ξ_k and $\tilde{\xi}_k$ which are obviously arbitrary. Therefore, from now on we choose $s_k = 0$ unless otherwise stated.

Exercise 4 Derive the relation

$$\Delta(k) = \left[\frac{i\mathcal{P}}{k_0 - \omega_k} + \pi\delta(k_0 - \omega_k)A_k(\theta) \right] \tag{7.132}$$

with

$$A_k(\theta) = B_k^{-1}(\theta)\tau_3 B_k(\theta) \tag{7.133}$$

$$= \left[\begin{array}{cc} 1 + 2\sigma n_k & -2f_k^{\alpha-1}n_k \\ 2\sigma f_k^{1-\alpha}[1 + \sigma n_k] & -[1 + 2\sigma n_k] \end{array} \right] \tag{7.134}$$

Here the symbol $\mathcal{P}$ means the principal value.

Hint: Make use of the formula

$$\frac{1}{\omega + i\epsilon_F} = \frac{\mathcal{P}}{\omega} - i\pi\delta(\omega). \tag{7.135}$$

It is useful to write the Bogoliubov matrix as a function of ω_k:

$$B_k(\theta) = B(\omega_k, \theta). \tag{7.136}$$

Then $A_k(\theta)$ is written as $A(\omega_k, \theta)$:

$$A(\omega, \theta) = B^{-1}(\omega, \theta)\tau_3 B(\omega, \theta). \tag{7.137}$$

We can then write Δ as

$$\Delta(k) = \left[\frac{i\mathcal{P}}{k_0 - \omega_k} + \pi\delta(k_0 - \omega_k)A(\omega_k, \theta) \right]. \tag{7.138}$$

Note that the thermal effect appears only through the matrix A. It is due to presence of the δ-function in the second term that we can replace ω_k by k_0:

$$\Delta(k) = \left[\frac{i\mathcal{P}}{k_0 - \omega_k} + \pi\delta(k_0 - \omega_k)A(k_0, \theta) \right]. \tag{7.139}$$

We now find the beautiful result

$$\Delta(k)^{\mu\nu} = [B^{-1}(k_0) \left(\frac{i}{k_0 - \omega_k + i\epsilon_F\tau_3} \right) B(k_0)]^{\mu\nu}. \tag{7.140}$$

Here, $B(k_0)$ means $B(k_0, \theta)$. Note that the argument of the Bogoliubov matrix is not ω_k but k_0, which is the Fourier form of time derivative.

Furthermore, when ω_k is non negative, we can replace ω_k by $|k_0|$:

$$\Delta(k)^{\mu\nu} = [B^{-1}(|k_0|)\left(\frac{i}{k_0 - \omega_k + i\epsilon_F\tau_3}\right)B(|k_0|)]^{\mu\nu}. \qquad (7.141)$$

This freedom in choice between k_0 and $|k_0|$ is frequently useful. A quite general form for the free propagator is obtained when we recall the rule (7.89):

$$B_{\alpha_1}(\omega) = W_{21}^{-1}(\omega)B_{\alpha_2}(\omega)W_{21}(\omega). \qquad (7.142)$$

Here W_{21} is treated as a function of ω_k:

$$W_{21}(\omega_k) = f^{\frac{(\alpha_2 - \alpha_1)}{2}\tau_3}(\omega_k). \qquad (7.143)$$

with f_k defined as a function of ω_k as $f_k = f(\omega_k)$.

Then we have

$$\Delta_{\alpha_1}(k)^{\mu\nu} = [W_{21}^{-1}(k_0)B_{\alpha_2}^{-1}(|k_0|)\left(\frac{i}{k_0 - \omega_k + i\epsilon_F\tau_3}\right)B_{\alpha_2}(|k_0|)W_{21}(k_0)]^{\mu\nu}, \qquad (7.144)$$

where the propagator with $\alpha = \alpha_1$ is denoted by $\Delta_{\alpha_1}(k)$. This relation is useful when we want to relate the propagators with different α, because this gives [2]

$$\Delta_{\alpha_1}(k) = W_{21}^{-1}(k_0)\Delta_{\alpha_2}(k)W_{21}(k_0). \qquad (7.145)$$

When the field equation for φ has the form

$$\left[\left(\frac{\partial}{\partial t}\right)^2 + \omega(-i\vec{\nabla})^2\right]\varphi(x) = 0, \qquad (7.146)$$

the field φ is said to be type 2. In this case we have two sets of oscillator operators; a_k and b_k. Their vacuum operators are denoted by $\xi_k(\theta)$ and $\eta_k(\theta)$, respectively. Here we consider a boson field.

Then we have

$$\varphi(x)^{\mu} = \int \frac{d^3k}{(2\pi)^{3/2}\sqrt{2\omega_k}}[e^{i[\vec{k}\cdot\vec{x} - \omega_k t]}a_k^{\mu} + e^{-i[\vec{k}\cdot\vec{x} - \omega_k t]}b_k^{\dagger\,\mu}] \qquad (7.147)$$

$$\bar{\varphi}(x)^{\mu} = \int \frac{d^3k}{(2\pi)^{3/2}\sqrt{2\omega_k}}[e^{-i[\vec{k}\cdot\vec{x} - \omega_k t]}\bar{a}_k^{\mu} + e^{i[\vec{k}\cdot\vec{x} - \omega_k t]}\bar{b}_k^{\dagger\,\mu}] \qquad (7.148)$$

with the relations:

$$b_k^{\mu} = B_k^{-1}(\theta)^{\mu\nu}\eta_k(\theta)^{\nu}, \quad \bar{b}_k^{\mu} = \bar{\eta}(\theta)^{\nu}B_k(\theta)^{\nu\mu}. \qquad (7.149)$$

Introducing the two fields

$$\varphi_a(x) = \int \frac{d^3k}{(2\pi)^{3/2}\sqrt{2\omega_k}} e^{i[\vec{k}\cdot\vec{x}-\omega_k t]} a_k, \tag{7.150}$$

$$\varphi_b(x) = \int \frac{d^3k}{(2\pi)^{3/2}\sqrt{2\omega_k}} e^{i[\vec{k}\cdot\vec{x}-\omega_k t]} b_k, \tag{7.151}$$

and making use of the thermal doublet notation, we can write them as

$$\varphi(x)^\mu = \varphi_a(x)^\mu + [\tau_3\bar{\varphi}_b^t(x)]^\mu, \tag{7.152}$$

$$\bar{\varphi}(x)^\mu = \bar{\varphi}_a^\mu + [\varphi_b^t(x)\tau_3]^\mu. \tag{7.153}$$

Here, the superscript t means transposition.

We now have

$$\langle 0(\theta)|T[\varphi(x)^\mu\bar{\varphi}(y)^\nu]|0(\theta)\rangle = \langle 0(\theta)|T[\varphi_a(x)^\mu\bar{\varphi}_a(y)^\nu]|0(\theta)\rangle$$
$$+ \left(\tau_3\{\langle 0(\theta)|T[\varphi_b(y)\bar{\varphi}_b(x)]|0(\theta)\rangle\}^t\tau_3\right)^{\mu\nu}. \tag{7.154}$$

Note that in the second term x and y are exchanged.

By applying the results obtained above for fields of type 1 to both φ_a and φ_b, we obtain the following causal propagator:

$$\Delta(x-y)^{\mu\nu} \equiv \langle 0(\theta)|T[\varphi(x)^\mu\bar{\varphi}(y)^\nu]|0(\theta)\rangle \tag{7.155}$$

$$= \frac{1}{(2\pi)^4}\int d^4k\, e^{ik(x-y)}\Delta(k)^{\mu\nu} \tag{7.156}$$

with

$$\Delta(k)^{\mu\nu} = \frac{i}{2\omega_k}\left[B^{-1}(k_0)\frac{1}{k_0-\omega_k+i\epsilon_F\tau_3}B(k_0)\right.$$
$$\left. - \tau_3 B^t(-k_0)\frac{1}{k_0+\omega_k-i\epsilon_F\tau_3}B^{-1t}(-k_0)\tau_3\right]^{\mu\nu}. \tag{7.157}$$

Use of the relation (7.142) then gives

$$\Delta_1(k)^{\mu\nu} = \frac{i}{2\omega_k}\left[W_{21}^{-1}(k_0)B_2^{-1}(k_0)\frac{1}{k_0-\omega_k+i\epsilon_F\tau_3}B_2(k_0)W_{21}(k_0)\right.$$
$$\left. - W_{21}(-k_0)\tau_3 B_2^t(-k_0)\frac{1}{k_0+\omega_k-i\epsilon_F\tau_3}B_2^{-1t}(-k_0)\tau_3 W_{21}^{-1}(-k_0)\right]^{\mu\nu}. \tag{7.158}$$

Here Δ_1 is the propagator with $\alpha = \alpha_1$.

When ω_k is non negative, we can write Δ as

$$\Delta_1(k)^{\mu\nu} = \frac{i}{2\omega_k} \left[W_{21}^{-1}(k_0)B_2^{-1}(|k_0|)\frac{1}{k_0 - \omega_k + i\epsilon_F\tau_3}B_2(|k_0|)W_{21}(k_0) \right.$$
$$\left. - W_{21}(-k_0)\tau_3 B_2^t(|k_0|)\frac{1}{k_0 + \omega_k - i\epsilon_F\tau_3}B_2^{-1t}(|k_0|)\tau_3 W_{21}^{-1}(-k_0) \right]^{\mu\nu} .$$
$$(7.159)$$

Now we choose $\alpha_2 = 1/2$, because then $B_2^{-1} = \tau_3 B_2^t\tau_3$. We obtain

$$\Delta_1(k)^{\mu\nu} = \frac{i}{2\omega_k} \left[W_{21}^{-1}(k_0)B_2^{-1}(|k_0|)\frac{1}{k_0 - \omega_k + i\epsilon_F\tau_3}B_2(|k_0|)W_{21}(k_0) \right.$$
$$\left. - W_{21}(-k_0)B_2^{-1}(|k_0|)\frac{1}{k_0 + \omega_k - i\epsilon_F\tau_3}B_2(|k_0|)W_{21}^{-1}(-k_0) \right]^{\mu\nu} .$$
$$(7.160)$$

for B_2 with $\alpha_2 = 1/2$.

These results take a particularly simple form when $\alpha_1 = 1/2$ is used, because then $\alpha_1 = \alpha_2$ which gives $W_{21} = 1$:

$$\Delta(k)^{\mu\nu} = [B^{-1}(|k_0|)\frac{i}{k_0^2 - \omega_k^2 + i\epsilon_F\tau_3}B(|k_0|)]^{\mu\nu}. \qquad (7.161)$$

Note that, since we are considering a thermal theory (even in a relativistic model the Lorentz symmetry is not respected), we do not assume the dispersion relation $\omega^2(\vec{k}) = m^2+\vec{k}^2$; the parameter $\omega(\vec{k})$ is still an arbitrary function of $\vec{k}$.

The above consideration for bosons of type 2 can easily be extended to fermions of type 2 such as the Dirac field.

The propagators studied in this subsection are the ones to be used when the Feynman diagram method is employed. When n_k assumes the equilibrium form, we find a simple form of the propagator for fields of type 2. This will be shown at the end of this section.

We now introduce an important theorem which states that when a quantum field model is renormalizable at zero temperature in the sense that all of the ultraviolet divergences are eliminated by the renormalization method, it is also renormalizable at finite temperature. To prove this [3], we rewrite the relation (7.139) as follows:

$$\Delta(k) = \left[\frac{i}{k_0 - \omega_k + i\epsilon_F\tau_3} + \pi\delta(k_0 - \omega_k)(A_k(k_0, \theta) - \tau_3) \right], \qquad (7.162)$$

where the matrix A_k was given in (7.134). Note that all the thermal effects are carried by the second term which exponentially vanishes at high energy limit, implying that the thermal effects do not create divergences worse than those at zero temperature. Making this splitting in each Feynman line and expanding a diagram in terms of power of the second term, we obtain a low temperature expansion.

When the number density n assumes the equilibrium form:

$$n(\omega) = \frac{1}{e^{\beta\omega} - \sigma}, \tag{7.163}$$

the expression (7.157) for the propagator of the field of type 2 can be further simplified. This is because the equilibrium n_k satisfies

$$1 + \sigma n(-k_0) = -\sigma n(k_0), \tag{7.164}$$

which gives

$$1 + 2\sigma n(-k_0) = -[1 + 2\sigma n(k_0)]. \tag{7.165}$$

Now recall (7.134) for the matrix $A = B^{-1}\tau_3 B$. This together with the above relations for $n(k_0)$ gives

$$A(-k_0) = -\tau_3 A^t(k_0)\tau_3, \tag{7.166}$$

which means

$$\tau_3 B^t(-k_0)\tau_3 B^{-1,t}(-k_0)\tau_3 = -B^{-1}(k_0)\tau_3 B(k_0). \tag{7.167}$$

With this relation we rewrite (7.157) as

$$\Delta(k)^{\mu\nu} = B^{-1}(k_0)\frac{i}{(k_0 + i\epsilon_F\tau_3)^2 - \omega_k^2}B(k_0). \tag{7.168}$$

Note that the (11)-component of this propagator is the Dolan-Jackiw propagator presented in the celebrated paper [4].

7.3   The General Structure of TFD

In the last section we explained how doubling the degrees of freedom in TFD can take care of thermal effects. Though our argument started with the study of the two mode squeezed state, the idea of doubling degrees of freedom to formulate an operator formalism for thermal theory for a quantum field system was much older. In subsection 5.3.6 it was shown that the Coulomb potential effect on the NG mode in superconductivity can be easily taken care of by a canonical transformation method. However, it is not easy to extend this calculation to a situation at finite temperature by means of thermal Green function formalisms, because it needs an operator formalism at finite temperature. This was the motivation behind the development in [5] of the thermo field dynamics with its doubled degrees of freedom. The fact that the particle number average at temperature

$T = 1/\beta$ is given by $n_k = f_k/[1 - \sigma f_k]$ with $f_k = \exp[-\beta\omega_k]$ suggested that these particles are condensed in the vacuum. It was then found that to cover such temperature freedom in particle number is easily provided by doubled degrees of freedom which introduce the freedom of the Bogoliubov transformation. Although the above paper [5] was written for a study of superconductivity, the TFD computational method including both operator method and Green's function method were developed sufficiently to perform most of the calculations required by the program of the paper. However, since the main emphasis in the paper was on superconductivity, it was not recognized as a general real time finite temperature formalism. The first attempt to put it as a systematically organized form of real time finite temperature theory was made in [6], where the theory was called thermo field dynamics (TFD). There the equivalence between the pure state vacuum expectation value and mixed state ensemble average (given in chapter 2 in this book) was explicitly shown. The thermal state condition was also presented. A systematic list of the tilde conjugation rules was presented in [7]. A summary of TFD and its applications up to 1982 are found in [8].

In this section we present the basis of TFD in a general form. In TFD every operator, say A, has a partner $\tilde{A}$ given according to the tilde conjugation rules given in [8, 1]. These rules are summarized here again:

$$(AB)^\sim = \tilde{A}\tilde{B}, \tag{7.169}$$
$$(c_1 A + c_2 B)^\sim = c_1^* \tilde{A} + c_2^* \tilde{B}, \tag{7.170}$$
$$(A^\dagger)^\sim = \tilde{A}^\dagger, \tag{7.171}$$
$$(\tilde{A})^\sim = \sigma A, \tag{7.172}$$
$$|0(\theta)\rangle^\sim = |0(\theta)\rangle, \tag{7.173}$$
$$\langle 0(\theta)|^\sim = \langle 0(\theta)|, \tag{7.174}$$

where c_1 and c_2 are any two c-numbers and A and B stand for any two operators.

When a system of quantum fields is given, its Lagrangian gives rise to the dynamical Hamiltonian H which is a Heisenberg operator. In TFD, H is not the Hamiltonian, because the Hamiltonian should generate the time-translation of both nontilde and tilde operators. We have denoted this Hamiltonian by $\hat{H}$. Since thermal effects should not change the Heisenberg equations for nontilde fields, any term containing both tilde and nontilde fields is forbidden in the Hamiltonian $\hat{H}$.

The Heisenberg equation for a Heisenberg field $\psi(x)$ has the form

$$i\frac{\partial}{\partial t}\psi(x) = [\psi(x), \hat{H}]. \tag{7.175}$$

Then, since any c-number becomes its complex conjugate under tilde conjugation, taking the tilde conjugate of this Heisenberg equation leads to the following Heisenberg equation for the tilde fields:

$$-i\frac{\partial}{\partial t}\tilde{\psi}(x) = [\tilde{\psi}(x), (\hat{H})^{\tilde{}}]. \tag{7.176}$$

This gives

$$(\hat{H})^{\tilde{}} = -\hat{H}. \tag{7.177}$$

These considerations lead to the form of $\hat{H}$

$$\hat{H} = H - \tilde{H}. \tag{7.178}$$

The above list of tilde conjugation rules and (7.178) for the structure of Hamiltonian form the general basis of TFD.

The eigenvalue of $\hat{H}$ will be called the hat-energy denoted by $\hat{E}$, while the expectation value of H is the dynamical energy E. Let us study those thermal vacua which are eigenvectors of $\hat{H}$. Thus we write

$$\hat{H}|0\rangle = \hat{E}|0\rangle. \tag{7.179}$$

Since $\hat{H}$, $|0\rangle$ and $\hat{E}$ transform under the tilde conjugation as

$$
\begin{aligned}
(\hat{H})^{\tilde{}} &= -\hat{H} \\
|0\rangle^{\tilde{}} &= |0\rangle \\
(\hat{E})^{\tilde{}} &= \hat{E},
\end{aligned}
\tag{7.180}
$$

we have from (7.179) the relation

$$\hat{E} = 0. \tag{7.181}$$

Thus all of the thermal vacua which are hat-energy eigenstates form a set of zero hat-energy states. Parameterizing these states by θ, we write the set as $\{|0(\theta)\rangle\}$. The negative sign carried by $\tilde{H}$ in (7.178) indicates that the set $\{|0(\theta)\rangle\}$ is continuous. The same negative sign causes another significant result, that is, the hat-energy spectrum is lower unbounded.

The relation (7.181) shows that

$$\hat{H}|0(\theta)\rangle = 0, \tag{7.182}$$

implying that each member of the set $\{|0(\theta)\rangle\}$ is a stationary state. This does not mean that time dependent thermal processes cannot be treated by TFD. As we will see in chapter 9, TFD seems to provide a good method for dealing with time dependent thermal physics. Here we only point out that the initial vacuum in a time-dependent thermal process is not an eigenvector of $\hat{H}$.

From the continuous set of zero hat-energy vacua there emerges the operation of zero hat-energy modes. Let us introduce the following operation $\hat{G}$ which drives a ket-state from one thermal vacuum to another:

$$|0(\theta + d\theta)\rangle = e^{id\theta\hat{G}}|0(\theta)\rangle \tag{7.183}$$

with infinitesimal $d\theta$; (7.183) derives for a bra-state

$$\langle 0(\theta + d\theta)| = \langle 0(\theta)|e^{-id\theta\hat{G}}, \tag{7.184}$$

since the vacuum should be normalized,

$$\langle 0(\theta)|0(\theta)\rangle = 1. \tag{7.185}$$

In (7.183) and (7.184), the product $d\theta\hat{G}$ is generally to be interpreted as the inner product:

$$d\theta\hat{G} = \sum_i d\theta_i \hat{G}_i. \tag{7.186}$$

with multi generators $\hat{G}_i$.

Since $\hat{G}$ is the zero hat-energy mode, one can always choose $\hat{G}$ in such a manner that it commutes with $\hat{H}$:

$$[\hat{H}, \hat{G}] = 0. \tag{7.187}$$

This implies that the Hamiltonian is invariant under the transformation generated by $\hat{G}$ ($\hat{G}$-symmetry).

However the vacuum is not invariant under this transformation:

$$\hat{G}|0(\theta)\rangle \neq 0 \tag{7.188}$$

and/or

$$\langle 0(\theta)|\hat{G} \neq 0. \tag{7.189}$$

Thus we see that the $\hat{G}$-symmetry is spontaneously broken and also that $\hat{G}$ acts as the operator of zero hat-energy mode, that is, the NG boson operator. The invariance of $|0(\theta)\rangle$ under the tilde conjugation leads to

$$(\hat{G})^{\sim} = -\hat{G}. \tag{7.190}$$

A summary of the above ideas is that in TFD the thermal degree of freedom is the freedom of moving through the zero hat-energy degenerate states and that the spontaneous breakdown of the $\hat{G}$-symmetry is associated with these hat-energy degenerate vacua. This is the TFD view of the thermal degree of freedom. It is obvious that this freedom does not exist in a quantum mechanical isolated system but that requires a quantum field system.

The formalism presented in this section is the basic form of TFD [9] upon which all future developments are based.

7.4   Thermal Observable

It has been repeated several times that the operators for dynamical observables in the Heisenberg representation are the nontilde operators. From this it has frequently been stated elsewhere that tilde operators are auxiliary operators which have nothing to do with real observations and it is due to this reason that tilde operators were sometimes called ghost operators. However, it becomes clear that this view is not correct when we recall that not all observables are dynamical; there are thermal observables such as the heat energy, entropy and so on.

The thermal observables are observed when a thermal situation changes. For example the operator for the dynamical energy is not the Hamiltonian $\hat{H}$, but H. Since the vacua $|0(\theta)\rangle$ are not eigenvectors of H, its vacuum expectation value changes when the thermal situation changes. In other words the dynamical energy expectation values do depend on θ. Thus, it requires a change of dynamical energy to move from one member to another in the set $\{|0(\theta)\rangle\}$ which is degenerate in the hat-energy. This change of dynamical energy is identified as the heat energy according to the law of energy conservation.

To make this argument concrete let us recall a thermodynamical relation between internal energy, heat energy and work. When we consider an isolated system of quantum fields, there is no external work. The internal energy is the dynamical energy mentioned above. We thus have

$$d\mathcal{E} = dQ. \tag{7.191}$$

Here $\mathcal{E}$ and Q are the dynamical energy density and heat energy density, respectively. They depend on θ as $\mathcal{E} = \mathcal{E}(\theta)$ and $Q(\theta)$. To illustrate the dynamical energy and heat energy we consider a free field whose Hamiltonian is

$$\hat{H}_0 = H_0 - \tilde{H}_0 \tag{7.192}$$

with

$$H_0 = \int d^3k \, \omega_k \, a_k^\dagger a_k. \tag{7.193}$$

Then, the dynamical energy density is obtained from

$$\mathcal{E}(\theta)[\delta(\vec{q})]_{\vec{q}=0} = \langle 0(\theta)|H_0|0(\theta)\rangle, \tag{7.194}$$

$$= [\delta(\vec{q})]_{\vec{q}=0} \int d^3k \, \omega_k \, n_k(\theta). \tag{7.195}$$

When θ changes, the change of dynamical energy density is

$$d\mathcal{E}(\theta) = \int d^3k \, \omega_k \, dn_k(\theta). \tag{7.196}$$

Introducing the usual form of entropy density

$$S(\theta) = \int d^3k \, [\sigma(1 + \sigma n_k(\theta)) \ln(1 + \sigma n_k(\theta)) - n_k(\theta) \ln n_k(\theta)], \tag{7.197}$$

with the unit $k_B = 1$, where k_B is the Boltzmann constant, we obtain

$$dS(\theta) = \int d^3k \, c_k(\theta) \, dn_k(\theta), \qquad (7.198)$$

where c_k is defined by

$$c_k(\theta) = -\ln f_k(\theta). \qquad (7.199)$$

The f_k is related to n_k as

$$n_k = \frac{f_k}{1 - \sigma f_k}. \qquad (7.200)$$

At temperature T, we have $c_k = \beta \omega_k$ with $\beta = 1/T$. Then, the relation (7.191) gives the well known thermodynamical relation

$$dS(\theta) = \frac{dQ(\theta)}{T} \qquad (7.201)$$

for an equilibrium situation [10]. In TFD the concept of temperature follows from the stability of the vacuum, as will be shown later in the next section.

Summarizing, the tilde degrees of freedom are needed to describe thermal observables. This provides us with the hope that TFD might open a way to incorporate thermodynamics into quantum field physics. However, this goal is still far from being achieved.

7.5   Perturbative Calculations

7.5.1   THE INTERACTION REPRESENTATION

Given the TFD formalism above, we now study perturbative calculations in TFD. To do this we need the relation between the Heisenberg and interaction representations [2].

In TFD one starts off with the full Hamiltonian density in the Heisenberg representation

$$\hat{h}_H(x) = h_H(x) - \tilde{h}_H(x). \qquad (7.202)$$

The spatial integration of this operator is the Hamiltonian in the Heisenberg representation $\hat{H}_H$. Throughout, the subscript H will denote a quantity in the Heisenberg representation. The $h_H(x)$ consists of the free part $h_{0H}(x)$ and the interaction part $h_{IH}(x)$:

$$h_H(x) = h_{0H}(x) + h_{IH}(x). \qquad (7.203)$$

Thus, the free Hamiltonian density and interaction Hamiltonian density are respectively given by

$$\hat{h}_{0H}(x) = h_{0H}(x) - \tilde{h}_{0H}(x), \qquad (7.204)$$
$$\hat{h}_{IH}(x) = h_{IH}(x) - \tilde{h}_{IH}(x). \qquad (7.205)$$

We then relate any operator $O_H(x)$ in the Heisenberg representation to the operator $O(x)$ in the interaction representation (unsubscripted quantities will be in the interaction representation) by

$$O_H(x) = \hat{U}^{-1}(t, t_0)\, O(x)\, \hat{U}(t, t_0), \tag{7.206}$$

where $t = t_x$ and

$$\frac{\partial}{\partial t}\hat{U}(t, t_0) = -i\hat{H}_I(t)\hat{U}(t, t_0), \tag{7.207}$$

Here $\hat{H}_I(t)$ is the interaction Hamiltonian in the interaction representation:

$$\hat{H}_I(t) = \int d^3x\, \hat{h}_I(x) \tag{7.208}$$

with

$$\hat{h}_I(x) = h_I(x) - \tilde{h}_I(x). \tag{7.209}$$

On the other hand, the free Hamiltonian in the interaction representation is denoted by $\hat{H}_0$. The structure of this free Hamiltonian was given in the previous section for free field TFD:

$$\hat{H}_0 = H_0 - \tilde{H}_0. \tag{7.210}$$

We then have

$$\frac{\partial}{\partial t}O(t) = -i[O(t), \hat{H}_0]. \tag{7.211}$$

Here we have chosen the Heisenberg and interaction representations to coincide at $t = t_0$: $\hat{U}(t_0, t_0) = 1$. Mathematically speaking, we need an elaboration in defining the $\hat{U}$-operator (for example, the use of an adiabatic factor in the coupling constant), because otherwise the interaction representation does not exist. We do not touch this point here.

The one body propagator in the Heisenberg representation is defined by

$$D(x - y) \equiv -i_H\langle 0(\theta)|T\left[\Phi_H(x)\, \Phi_H(y)\right]|0(\theta)\rangle_H. \tag{7.212}$$

The fields $\Phi_H(x)$ and $\Phi_H(y)$ denote any tilde or nontilde fields, not necessarily the same.

Now we write (7.212) in terms of operators in the interaction representation. Identifying the Heisenberg vacua with the interaction vacua at t_0,

$$|0(\theta)\rangle_H = |0(t_0, \theta)\rangle, \tag{7.213}$$
$$_H\langle 0(\theta)| = \langle 0(t_0, \theta)|. \tag{7.214}$$

we find

$$D(x - x') = -i\langle 0(t_0, \theta)|\hat{U}(t_0, \infty)T[\Phi(x)\, \Phi(x')\, \hat{S}]\hat{U}(-\infty, t_0)|0(t_0, \theta)\rangle \tag{7.215}$$

Here $\hat{S}$ is given by

$$\hat{S} = \hat{U}(\infty, -\infty), \qquad (7.216)$$

which is called the $\hat{S}$-operator. This should not be confused with the S-matrix operator which relates in-fields to out-fields, because there are no stable asymptotic particles in thermal situations, contrary to the situation found at zero-temperature. This will be discussed in the next chapter.

At this stage in ordinary quantum field theory one usually introduces the Gell-Mann Low formula to eliminate $\hat{U}(t_0, \infty)$ and $\hat{U}(-\infty, t_0)$, and obtains the usual time ordered propagators in the interaction representation that are amenable to a Feynman diagram expansion. However, since the Gell-Mann Low formula may not be valid in arbitrary thermal situations we avoid its use, and find an alternate method.

Note that usability of the Feynman diagram method is a matter of convenience. Even when $\hat{U}(t_0, \infty)$ and $\hat{U}(-\infty, t_0)$ are not eliminated, one can calculate (7.215) in terms of retarded and advanced propagators, though the calculation becomes very clumsy. It is therefore useful to study how and when one can derive a time ordered formalism *without using the Gell-Mann Low formula*.

To do this we choose $t_0 = -\infty$ or $t_0 = \infty$. Since t_0 is specified, we do not explicitly write t_0 in the thermal vacuum. We then have

$$D(x - y) = -i\langle 0(\theta)|\hat{S}^{-1}T[\Phi(x)\,\Phi(y)\,\hat{S}]|0(\theta)\rangle \qquad (7.217)$$

for $t_0 = -\infty$, and

$$D(x - y) = -i\langle 0(\theta)|T[\Phi(x)\,\Phi(y)\,\hat{S}]\hat{S}^{-1}|0(\theta)\rangle \qquad (7.218)$$

for $t_0 = \infty$.

In the following we are mostly concerned with the choice $t_0 = -\infty$. We have then got rid of $\hat{U}(-\infty, t_0)$. However, the presence of $\hat{S}$ outside the T-bracket in (7.217) still prevents us from using the Feynman diagram method. There are two ways to eliminate this $\hat{S}$.

One way is to choose $\alpha = 1$. The relation (7.112) in subsection 7.2.5 shows that

$$\langle 0(\theta)|\hat{H}_I(t) = 0 \quad \text{for } \alpha = 1. \qquad (7.219)$$

which gives

$$\langle 0(\theta)|\hat{S}^{-1} = \langle 0(\theta)| \quad \text{for } \alpha = 1. \qquad (7.220)$$

With this we can make use of the Feynman diagram method.

The thermal state condition in subsection 7.2.5 also gives

$$\hat{H}_I(t)|0(\theta)\rangle = 0 \quad \text{for } \alpha = 0. \qquad (7.221)$$

which gives

$$\hat{S}^{-1}|0(\theta)\rangle = |0(\theta)\rangle \quad \text{for } \alpha = 0. \qquad (7.222)$$

This shows that $\alpha = 0$ with $t_0 = \infty$ also gives the purely time ordered expression which makes the Feynman diagram method available to us.

This method of using particular choices of α is applicable to any form of number density n_k.

However, this does not mean that the α-independence of vacuum expectation values of dynamical (i.e. non tilde) operators is lost. With any value of α, we can calculate the vacuum expectation values of dynamical observable such as the (11)-component of the two-point propagator studied above. The above conclusion is that for arbitrary n_k the calculation requires a computation more complicated than the Feynman method unless the particular values of α are chosen. However, when we have a stable vacuum, the purely chronological method (i.e. the Feynman method) is usable for any α. We will study this in the next subsection.

It should be noted that $\hat{S}$ itself is given by ordered products of interaction, and therefore, can always be computed by the Feynman diagram method. However, it is not a dynamical operator, and therefore, it can depend on α in a general situation.

7.5.2 VACUUM STABILITY AND TEMPERATURE STATES

The states of a system of quantum fields belong to a Fock space which is built on a vacuum. All the states except the vacuum, that is, the states with quasi-particles, are excited states. It is due to the thermal dissipation that all the excited states are unstable. However the thermal vacuum can be stable.

When a thermal situation is stationary it does not necessarily mean the stability of the vacuum; a stationary vacuum implies only that n_k is independent of time. Consider a system in which the supply of particles in a beam from the outside is continuous. This is one way to keep a time independent particle number. In such a situation the number n_k can assume any function of $\vec{k}$, depending on the external effect. This is the kind of situation discussed in the latter part of the last subsection. In that case the stability of the vacuum is achieved only when the interaction H_{IH} is modified to include the interaction between the system and the external beam in calculations of the $\hat{S}$-operator. The external effect is usually expressed by a c-number field and the present analysis requires a modification. When we include the external system as a quantum field, the spatial translation invariance is lost because the system and the source of the external effect occupy different locations. This also requires a modification of the present analysis.

However, when a system preserves a stationary number without an external particle supply, the stationary thermal situation may mean the stability of the vacuum; the vacuum supports the stationary thermal situation. Since the stable vacuum supports itself self-consistently, it may permit only a particular choice for n_k as a function of ω_k. This is an intuitively reasonable story, though it requires a mathematical derivation.

In this section we try to clarify this problem. Let us first ask how the stability of the vacuum is mathematically formulated. Any short time measurement induces energy uncertainty, causing all kinds of quantum fluctuation. Therefore, as is well known in the usual quantum field theory, the stability of the vacuum is tested with measurements over an infinite time. This situation is not changed by thermal effects. The stability of the vacuum is therefore expressed by the following relations:

$$\hat{S}|0(\theta)\rangle = \hat{S}^{-1}|0(\theta)\rangle = |0(\theta)\rangle, \tag{7.223}$$

$$\langle 0(\theta)|\hat{S} = \langle 0(\theta)|\hat{S}^{-1} = \langle 0(\theta)|. \tag{7.224}$$

In this self-consistent stationary situation we have a good reason to believe that the time ordered formulation (the Feynman diagram method) is workable. When Stückelberg [11] derived the Feynman formalism, it was the causality condition which led to the time ordered products. This causality condition dictates that an effect to a future point is given by a diverging wave starting from the present point, while an effect from the past is given by a wave converging to the present point. This classic work of Stückelberg suggests that the same principle be required for a self-consistent stable vacuum. Indeed, the stability condition given by (7.223) and (7.224) is just the condition which makes the Green's functions in (7.217) and (7.218) time ordered functions, permitting the Feynman diagram method.

We presume that this consideration is applicable to any choice of the parameter α. Indeed, when $\hat{S}$ is independent of α, it satisfies the above stability condition, because we have seen that (7.223) is satisfied by $\alpha = 0$ with any n_k, while (7.224) is satisfied by $\alpha = 1$. Thus our task now is to find a condition for α-independent $\hat{S}$.

We are ready to give the theorem for the operator $\hat{S}$ [2].

Theorem 1 *Each matrix element of the operator $\hat{S}$ in the Fock space built on the vacua $\langle 0(\theta)|$ and $|0(\theta)\rangle$ is independent of α provided that n_k is given by the canonical distribution; $n_k = f_k/[1 - \sigma f_k]$ with $f_k = \exp[-\beta\omega_k]$.*

To prove the theorem, we recall the T-product formula for $\hat{S}$,

$$\hat{S} = \sum_{n=0}^{\infty} \frac{(-i)^n}{n!} \int d^4x_1 \cdots d^4x_n \, T[\hat{h}_I(x_1) \cdots \hat{h}_I(x_n)]. \tag{7.225}$$

Since this consists only of the time ordered products, it is given by the Feynman diagrams with each internal line being given by the propagator $\Delta(k)$ presented in subsection 7.2.6. As it was shown in the same subsection, a free field of type 2 can be treated as two free fields of the type 1. Therefore, it is sufficient to prove the theorem for fields of the type 1.

The basic tool for the proof is the relation (7.145) in subsection 7.2.6, which transforms the one body propagator with α_2 to the one with α_1 by the matrix W_{21} which was given by

$$W_{21}(\omega_k) = f^{\frac{(\alpha_2 - \alpha_1)}{2}\tau_3}(\omega_k). \tag{7.226}$$

According to this rule to the Feynman diagrams with α_1, every φ carries the factor $f^{[(\alpha_2-\alpha_1)/2]}(k_0)$ added to the propagator with α_2, while every $\varphi^\dagger$ has the factor $f^{[-(\alpha_2-\alpha_1)/2]}(k_0)$.

A significant aspect of the vertices is that no vertex mixes nontilde and tilde fields, because the interaction Hamiltonian has no terms which mix nontilde with tilde fields.

We first focus our attention to a vertex attached to the internal lines only. Suppose a vertex consists of two φ and one $\varphi^\dagger$. This vertex creates the extra factor

$$F(k_0^1, k_0^2 : k_0^3) = \frac{f^{[(\alpha_2-\alpha_1)/2]}(k_0^1) f^{[(\alpha_2-\alpha_1)/2]}(k_0^2)}{f^{[(\alpha_2-\alpha_1)/2]}(k_0^3)}, \tag{7.227}$$

where k_0^1 and k_0^2 are the k_0 attached to two φ, while k_0^3 is the one attached to $\varphi^\dagger$. Since this factor appears in front of $\delta(k_0^1 + k_0^2 - k_0^3)$ which shows the energy conservation at each vertex, the α independence of the Feynman diagrams requires

$$F(k_0^1, k_0^2 : k_0^3)\delta(k_0^1 + k_0^2 - k_0^3) = \delta(k_0^1 + k_0^2 - k_0^3). \tag{7.228}$$

This implies that f has the form

$$f(k_0) = e^{-\beta k_0} \tag{7.229}$$

with a constant β. The constant $T \equiv 1/\beta$ is called the temperature. The number

$$n_k = n(k_0) = \frac{f(k_0)}{1 - \sigma f(k_0)}, \tag{7.230}$$

then has the equilibrium distribution with the temperature T.

The same argument is applied to vertices with tilde fields only.

Suppose that a line attached to a vertex is the external nontilde line. Since $B_{\alpha_1} = W_{21}^{-1} B_{\alpha_2} W_{21}$ according to subsection 7.2.6, the W_{21}^{-1} provides the same factor as the internal line. The W_{21} appears as a factor in matrix elements of $\hat{S}$. However, since the total energy associated with each matrix element is conserved, this extra factor contributed by all the external lines compensate in the same way as the compensation at each vertex, when $f(k_0)$ has the above equilibrium form. This completes the proof of theorem 1.

Since $\hat{S}$ is independent of α in an equilibrium situation, equilibrium vacua are stable, and all the multi point propagators can be given by the time ordered expansion (the Feynman expansion). This aspect of multi point propagators was proven in [12], in which the Feynman expansion was assumed, whereas the reliability of the Feynman expansion was questioned in [2]. This supplements the proof of α-independence.

Summarizing, it was shown that the sufficient condition for the stability of the vacuum for any value of α is the equilibrium number distribution. In this case the Feynman diagram method is applicable to any α. It was not proven that no other choice of n_k with a particular α gives a stable vacuum, though the complexity of integrations due to higher order loop correction suggests that this seems to be unlikely. The above derivation of temperature from stability of the vacuum may reflect equality between long time average and statistical average.

When n_k is independent of time but does not have the form of an equilibrium distribution, computation of propagators needs a method more complicated than the Feynman diagram method unless $\alpha = 1$ or $\alpha = 0$. This is a time independent nonequilibrium situation. In this case the choice $\alpha = 1$ or 0 is recommendable because of the ease of calculation. For example, when the Korenman model [13] for a nonequilibrium stationary problem was studied in the TFD framework in [14], the choice $\alpha = 1$ was used.

7.6 REFERENCES

[1] H. Umezawa and Y. Yamanaka, *Advances in Physics* **37** (1988) 531.

[2] T. S. Evans, I. Hardman, H. Umezawa and Y. Yamanaka, *J. Math. Phys.* **33** (1992) 370.

[3] H. Matsumoto, I. Ojima and H. Umezawa, *Ann. Phys.* **152** (1984) 348.

[4] L. Dolan and R. Jackiw, *Phys. Rev.* **D9** (1974) 3312.

[5] L. Leplae, H. Umezawa and F. Mancini, *Physics Reports* **10C** (1974) 153.

[6] Y. Takahashi and H. Umezawa, *Collect. Phenom.* **2** (1975) 55.

[7] H. Matsumoto, Y. Nakano and H. Umezawa, *Phys. Rev.* **D31** (1985) 429.

[8] H. Umezawa, H. Matsumoto and M. Tachiki, *Thermo Field Dynamics and Condensed States*, (North-Holland, Amsterdam, 1982).

[9] H. Umezawa and Y. Yamanaka, *Phys. Lett.* **155A** (1991) 75.

[10] I. Hardman, H. Umezawa and Y. Yamanaka, *Phys. Lett.* **146A** (1990) 293.

[11] E. C. G. Stueckelberg and T. A. Green, *Helv. Phys. Acta.* **24** (1951) 153.

[12] H. Matsumoto, Y. Nakano and H.Umezawa, *J. Math. Phys.* **25** (1984) 3076.

[13] V. Korenman, *Ann. Phys.* **39** (1969) 1387.

[14] I. Hardman, H. Umezawa and Y. Yamanaka, *Physica* **156A** (1989) 853.

8

Equilibrium TFD

8.1 Temporal Fourier Representation

8.1.1 THE CHOICES FOR α

Equilibrium TFD is characterized by the following choice for $f_k = f(\omega_k)$:

$$f(\omega_k) = e^{-\beta\omega_k}. \qquad (8.1)$$

Here, the quasi-particle energy ω_k includes the effects of renormalization due to interactions. This will be discussed in a later subsection. The definition of this energy frequently also includes the chemical potential. For example, in the case of the BCS model for superconductivity, the quasi electron energy has the form $\omega_k = [\epsilon_k^2 + \Delta^2]^{1/2}$ with the energy gap Δ; the ϵ_k is the normal electron energy which includes the chemical potential μ as $\epsilon_k = [(1/2m)|\vec{k}|^2 - \mu]$. Here $\mu = (1/2m)k_F^2$ with the Fermi momentum k_F. Since the treatment of chemical potential depends on the precise nature of a problem in many body systems, we do not touch this problem in this book, which looks at the general field theoretical aspects of many body theory.

As it was shown in subsection 7.5.2, the equilibrium vacua are stable for any choice of α, and the Feynman diagram method is applicable for any α.

Let us begin our consideration with the free field propagator, because this is the internal line in Feynman diagrams. Let us start with the free field of type 2. According to (7.168)

$$\Delta(k)^{\mu\nu} = B^{-1}(k_0)\frac{i}{(k_0 + i\epsilon_F\tau_3)^2 - \omega_k^2}B(k_0). \qquad (8.2)$$

According to (7.140) the propagator for fields of type 1 has the simple forms

$$\Delta(k)^{\mu\nu} = [B^{-1}(k_0)\left(\frac{i}{k_0 - \omega_k + i\epsilon_F\tau_3}\right)B(k_0)]^{\mu\nu} \qquad (8.3)$$

$$= [B^{-1}(|k_0|)\left(\frac{i}{k_0 - \omega_k + i\epsilon_F\tau_3}\right)B(|k_0|)]^{\mu\nu}. \qquad (8.4)$$

for any α.

The choice $\alpha = 1/2$ has merit in the sense that it presents TFD in a unitary representation. It provides us with the easiest way of inheriting the operator formalism of the usual quantum field theory. Originally, TFD was formulated with this choice of α, as was described in [1]. According to (7.88) the Bogoliubov matrix is

$$B_k^{-1}(\theta)^{\mu\nu} = (1 + \sigma n_k)^{1/2} \begin{bmatrix} 1 & f_k^{1/2} \\ \sigma f_k^{1/2} & 1 \end{bmatrix}, \qquad (8.5)$$

for $\alpha = 1/2$. Here we chose $s_k = 0$.

The choice $\alpha = 1$ has a different merit due to the fact that its thermal effects are contained entirely by the ket vacuum. This can be seen from the thermal sate condition (7.112) in subsection 7.2.5, which with $\alpha = 1$ reads as

$$\langle 0(\theta) | [A - \tilde{A}^\dagger] = 0. \qquad (8.6)$$

This shows that the bra vacuum is independent of time, because this gives $\langle 0(\theta) | \hat{H} = 0$, using the fact that $\hat{H}$ is invariant under the dagger conjugation. According to (7.88) we have an extremely simple form for the Bogoliubov matrix:

$$B_k^{-1}(\theta)^{\mu\nu} = (1 + \sigma n_k)^{1/2} \begin{bmatrix} 1 & f_k \\ \sigma & 1 \end{bmatrix}, \qquad (8.7)$$

for $\alpha = 1$. Here too, we chose $s_k = 0$.

Up to recent times the choice $s_k = 0$ has been commonly used. However the new choice, that is $(\alpha = 1, s_k = \ln[1 + \sigma n_k]^{1/2})$ appears to be also extremely useful, because it gives

$$B_k = \begin{bmatrix} 1 + \sigma n_k & -n_k \\ -\sigma & 1 \end{bmatrix}, \qquad (8.8)$$

the inverse of which is

$$B_k^{-1} = \begin{bmatrix} 1 & n_k \\ \sigma & 1 + \sigma n_k \end{bmatrix}. \qquad (8.9)$$

The fact that these matrices are linear in n_k frequently simplifies calculations. This choice will be extensively used in the next chapter. Future development of TFD is expected to rely heavily on use of this choice.

Although the two choices $\alpha = 1$ or $\alpha = 1/2$ are the commonly used ones, all the arguments in the following sections are developed without specifying α, because results of calculations of physical observables are independent of α. Therefore, in this chapter the choice of α is arbitrary, unless otherwise stated.

8.1.2   PRODUCT RULES FOR LOOPS

Since we are considering an equilibrium situation, the number parameter has the canonical form:

$$n(\omega_k) = \frac{f(\omega_k)}{1 - \sigma f(\omega_k)} \qquad (8.10)$$

with

$$f(\omega_k) = e^{-\beta\omega_k}. \tag{8.11}$$

Then we have the following product rule:

$$-i \int \frac{dl_0}{2\pi} \left[\tau_3 B_{\sigma_1}^{-1}(\kappa_+) \left(l_0 + \frac{k_0}{2} - \kappa_+ + i\epsilon_F \tau_3 \right)^{-1} B_{\sigma_1}(\kappa_+) \right]^{\mu\nu}$$

$$\times \left[B_{\sigma_2}^{-1}(\kappa_-) \left(l_0 - \frac{k_0}{2} - \kappa_- + i\epsilon_F \tau_3 \right)^{-1} B_{\sigma_2}(\kappa_-) \right]^{\nu\mu}$$

$$= \int d\kappa \rho(\kappa : \kappa_+, \kappa_-) [B_{\sigma_1\sigma_2}^{-1}(\kappa)(k_0 - \kappa + i\epsilon_F \tau_3)^{-1} B_{\sigma_1\sigma_2}(\kappa)]^{\mu\nu},$$

$$\tag{8.12}$$

where

$$\rho(\kappa : \kappa_+, \kappa_-) = \delta(\kappa - \kappa_+ + \kappa_-)\sigma[n(\kappa_+) - n(\kappa_-)]. \tag{8.13}$$

Here $B(\kappa)$ and $n(\kappa)$ are $B(\omega_k)$ and $n(\omega_k)$ with ω_k being replaced by κ. The suffix σ of B_σ means $\pm$; B_+ means B for boson field, while B_- is the B for fermion. Thus this product rule contains all of the possible combination of boson and fermion lines.

Exercise 1 Derive the above product rule.

Hint: Using the rule for change of α discussed in the last section, prove that this formula holds true for any α. Then we can choose any particular α such as $\alpha = 1$ or $\alpha = 1/2$. Choose for example $\alpha = 1/2$. Beyond this there are no more hints; simply calculate the left hand side for each choice of $\mu\nu$.

This product rule means that a loop diagram consisting of two Feynman lines is a superposition of the free propagator integrated over the energy parameter. When a two point propagator has the form of superposition of the free propagator integrated over the energy parameter it is called the spectral representation of two point propagator. The above product rule shows that the loop diagram has the spectral representation.

The derivation of the above product rule teaches us that this formula does not hold true unless the number distribution has the equilibrium distribution. This proves that the spectral representation needs the presence of temperature. It can be also proven that the equilibrium distribution for $n(\omega_k)$ is a sufficient condition for the existence of the spectral representation for any two point propagator with interaction effects. This will be discussed in the next subsection.

Let us consider a one loop self-energy given by the quantity in (8.12). Then, κ_+ and κ_- are functions of $[\vec{k} + (\vec{l}/2)]$ and $[\vec{k} - (\vec{l}/2)]$ respectively. Here, $\vec{k}$ and $\vec{l}$ are the external momentum and relative momentum respectively. Let us denote the quantity in (8.12) in this case by $\sigma_1(k_0, \vec{k} : \vec{l})$; the suffix 1 means one loop. Then the one loop self-energy $\Sigma_1(k_0, \vec{k})$ is given by

$$\Sigma_1(k_0, \vec{k}) = \int \frac{d^3l}{(2\pi)^3} \sigma_1(k_0, \vec{k} : \vec{l}) \tag{8.14}$$

Its imaginary part is

$$\Im\Sigma_1(k_0, \vec{k}) = -i\pi \int \frac{d^3l}{(2\pi)^3} \rho(k_0 : \kappa_+, \kappa_-) A(k_0) \tag{8.15}$$

with

$$A(k_0) = B^{-1}(k_0)\tau_3 B(k_0). \tag{8.16}$$

Here B means $B_{\sigma_1 \sigma_2}$.

Exercise 2 Derive this result for the imaginary part of the one loop self-energy.

Hint: Make use of the formula

$$\frac{1}{\omega \pm i\epsilon_F} = \frac{\mathcal{P}}{\omega} \mp i\pi\delta(\omega). \tag{8.17}$$

The on-shell self-energy is then given by $\Re\Sigma_1(\omega_k, \vec{k})$. The imaginary part of the on-shell self-energy $\Im\Sigma_1(\omega_k, \vec{k})$ does not vanish when the condition $\omega_k = \kappa_+ - \kappa_-$ is satisfied for some values of internal momentum $\vec{l}$. When this happens, the quasi-particle becomes unstable. This relation is frequently satisfied. The origin of this instability is the fact that the tilde particle carries the negative hat-energy. This means that a quasi-particle is decaying into a positive hat-energy nontilde particle and a negative hat-energy tilde particle. In the next section, we show that all the quasi-particles are unstable when there exist field interactions.

8.1.3   THE SPECTRAL REPRESENTATION

In this subsection we derive the spectral representation for two point propagators including all the interaction effects. The basic tool is the thermal state conditions in 7.2.5.

In an equilibrium situation these thermal state conditions give

$$[A(x) - e^{\beta\alpha\hat{H}}\tilde{A}^\dagger(x)]||0(\theta)\rangle = 0, \tag{8.18}$$

$$\langle 0(\theta)|[A(x) - \tilde{A}^\dagger(x)e^{(1-\alpha)\beta\hat{H}}] = 0. \tag{8.19}$$

We begin our considerations with the choice $\alpha = 1/2$. The result will be extended to all choices of α.

Let us apply the Fourier transformation to $A(x)$:

$$A(x) = \int d\kappa A(\vec{x}, \kappa)e^{-i\kappa t}, \tag{8.20}$$

$$\tilde{A}(x) = \int d\kappa \tilde{A}(\vec{x}, \kappa)e^{i\kappa t}. \tag{8.21}$$

Here the tilde conjugation rule was used.

Since $\hat{H}$ is the generator of time translation, we have

$$[A(\vec{x}, \kappa), \hat{H}] = \kappa A(\vec{x}, \kappa), \tag{8.22}$$

which gives

$$e^{c\hat{H}} A(\vec{x}, \kappa) e^{-c\hat{H}} = e^{-c\kappa} A(\vec{x}, \kappa) \tag{8.23}$$

with any c-number c.

Then, the above thermal state conditions become

$$[A(\vec{x}, \kappa) - e^{-\frac{1}{2}\beta\kappa} \tilde{A}^{\dagger}(\vec{x}, \kappa)]|0(\theta)\rangle = 0, \tag{8.24}$$

$$\langle 0(\theta)|[A^{\dagger}(\vec{x}, \kappa) - e^{-\frac{1}{2}\beta\kappa} \tilde{A}(\vec{x}, \kappa)] = 0. \tag{8.25}$$

Here the relation $\langle 0(\theta)|\hat{H} = \hat{H}|0(\theta)\rangle = 0$ is made use of.

Now, define

$$\Xi_A(\vec{x} : \kappa) = (1 + \sigma n(\kappa))^{1/2} [A(\vec{x}, \kappa) - e^{-\frac{1}{2}\beta\kappa} \tilde{A}^{\dagger}(\vec{x}, \kappa)], \tag{8.26}$$

which gives

$$\Xi_A^{\dagger}(\vec{x} : \kappa) = (1 + \sigma n(\kappa))^{1/2} [A^{\dagger}(\vec{x}, \kappa) - e^{-\frac{1}{2}\beta\kappa} \tilde{A}(\vec{x}, \kappa)]. \tag{8.27}$$

Here, $n(\kappa)$ is the number parameter related to $f(\kappa) = \exp[-\beta\kappa]$ through $n(\kappa) = f(\kappa)/[1 - \sigma f(\kappa)]$.

We then have

$$\Xi_A(\vec{x} : \kappa)|0(\theta)\rangle = 0, \tag{8.28}$$

$$\langle 0(\theta)|\Xi_A^{\dagger}(\vec{x} : \kappa) = 0. \tag{8.29}$$

Tilde conjugation of these relations give

$$\tilde{\Xi}_A(\vec{x} : \kappa)|0(\theta)\rangle = 0, \tag{8.30}$$

$$\langle 0(\theta)|\tilde{\Xi}_A^{\dagger}(\vec{x} : \kappa) = 0. \tag{8.31}$$

These relations show that Ξ_A and $\tilde{\Xi}_A$ act like annihilation operators and $\Xi_A^{\dagger}$ and $\tilde{\Xi}_A^{\dagger}$ are creation operators, although they are not oscillator operators. We call these relations diagonal thermal state relations.

Making use of the thermal doublet notation, we introduce $\Xi_A(\vec{x} : \kappa)^{\mu}$ and $\bar{\Xi}_B(\vec{x} : \kappa)^{\mu}$. Here, Ξ_B means Ξ_A with A being replaced by B. The doublet notation is the same as defined previously: $\Xi^1 = \Xi, \Xi^2 = \tilde{\Xi}^{\dagger}, \bar{\Xi}^1 = \Xi^{\dagger}, \bar{\Xi}^2 = -\sigma\tilde{\Xi}$. We also apply the thermal doublet notation to A: $A^1 = A, A^2 = \tilde{A}^{\dagger}, \bar{A}^1 = A^{\dagger}, \bar{A}^2 = -\sigma\tilde{A}$.

With this notation we have

$$A(\vec{x}, \kappa)^{\mu} = B^{-1}(\kappa)^{\mu\nu} \Xi_A(\vec{x}, \kappa)^{\nu}, \tag{8.32}$$

$$\bar{A}(\vec{x}, \kappa)^{\mu} = \bar{\Xi}_A(\vec{x}, \kappa)^{\nu} B(\kappa)^{\nu\mu}, \tag{8.33}$$

where $B(\kappa)$ is the Bogoliubov matrix with the number parameter given by $n(\kappa)$.

Exercise 3 Derive the following formula:

$$\langle 0(\theta)|T[\Xi_A(\vec{x},\kappa)^\mu \bar{\Xi}_B(\vec{y},\kappa)^\nu]|0(\theta)\rangle$$

$$= \rho(\vec{x} - \vec{y}, \kappa) \begin{bmatrix} \theta(t_x - t_y) & 0 \\ 0 & -\theta(t_y - t_x) \end{bmatrix}. \qquad (8.34)$$

Here

$$\rho(\vec{x} - \vec{y}, \kappa) \equiv \langle 0(\theta)|\Xi_A(\vec{x},\kappa)\Xi_B^\dagger(\vec{y},\kappa)|0(\theta)\rangle. \qquad (8.35)$$

Hints: Make use of the above definition of thermal doublets and the diagonal thermal state conditions. We also need the relation

$$\langle 0(\theta)|\Xi_A(\vec{x},\kappa)\Xi_B^\dagger(\vec{y},\kappa)|0(\theta)\rangle = \langle 0(\theta)|\tilde{\Xi}_B(\vec{y},\kappa)\tilde{\Xi}_A^\dagger(\vec{x},\kappa)|0(\theta)\rangle. \qquad (8.36)$$

This is obtained by means of a combination of dagger conjugation and tilde conjugation.

Now recall the relations

$$\theta(t_x - t_y)e^{-i\kappa(t_x - t_y)} = i \int \frac{dk_0}{2\pi} e^{-ik_0(t_x - t_y)} \frac{1}{k_0 - \kappa + i\epsilon_F}, \qquad (8.37)$$

$$\theta(t_y - t_x)e^{-i\kappa(t_x - t_y)} = -i \int \frac{dk_0}{2\pi} e^{-ik_0(t_x - t_y)} \frac{1}{k_0 - \kappa - i\epsilon_F} \qquad (8.38)$$

When we define the propagator

$$D_{AB}(x - y)^{\mu\nu} \equiv -i\langle 0(\theta)|T[A(x)^\mu \bar{B}(y)^\nu]|0(\theta)\rangle \qquad (8.39)$$

and recall that A^μ (B^μ) is related to Ξ_A^μ (Ξ_B^μ) through the Bogoliubov transformation in (8.32) and (8.33), the Fourier amplitude of this propagator has the spectral representation [1]:

$$D_{AB}(k)^{\mu\nu} = \int_{-\infty}^{\infty} d\kappa \rho_k(\kappa) \left[B^{-1}(\kappa) \frac{1}{k_0 - \kappa + i\epsilon\tau_3} B(\kappa) \right]^{\mu\nu}. \qquad (8.40)$$

Here $\rho_k(\kappa)$ is the Fourier component of $\rho(\vec{x} - \vec{y} : \kappa)$. This is called the spectral function.

In this derivation of the spectral representation the equilibrium condition $f(\kappa) = \exp[-\beta\kappa]$ played a decisive role. This shows that the necessary and sufficient condition for the existence of the spectral representation is that the number parameter has the equilibrium distribution. The existence of the spectral representation in equilibrium TFD is very useful in the calculation of the Feynman diagrams.

Although the choice $\alpha = 1/2$ was used in the above derivation of the spectral representation, the spectral representation holds true for any choice of α.

When we choose $A = B = \psi$, this propagator is denoted by $D_c(k)$. This is the propagator of the ψ field corrected by interaction effects. Now recall the Dyson equation

$$D_c = -i\Delta_c + (-i\Delta_c)\Sigma D_c, \tag{8.41}$$

where $\Sigma(k)$ is the self-energy term and Δ_c is the free field propagator.

Because the propagator has the spectral representation (8.40), the self-energy diagram gives a term of a similar spectral form:

$$\Sigma(k_0, \vec{k} : \beta)^{\mu\nu} = \int d\kappa\, \sigma_k(\kappa) \left[B^{-1}(\kappa) \frac{1}{k_0 - \kappa + i\epsilon\tau_3} B(\kappa) \right]^{\mu\nu}. \tag{8.42}$$

The spectral function, σ_k, of the self-energy is related to the spectral function, ρ_k, of the propagator in (8.40) through

$$\rho_k(\kappa) = \frac{\sigma_k(\kappa)}{\{\kappa - \omega_k - r_k(\kappa)\}^2 + \pi^2\sigma_k^2(\kappa)} \tag{8.43}$$

where

$$r_k(\kappa) = \int d\nu\, P \frac{\sigma_k(\nu)}{\kappa - \nu}. \tag{8.44}$$

Note that $r_k(k_0)$ is the real part of the self-energy:

$$r_k(k_0) = \Re\left[\Sigma\left(k_0, \vec{k} : \beta\right)\right]. \tag{8.45}$$

The on-shell self-energy, $\delta\omega_k(\beta)$, is obtained from the real part of Σ with $k_0 = \omega_k(\beta)$:

$$\delta\omega_k(\beta) = r_k(\omega_k(\beta)). \tag{8.46}$$

We usually choose ω_k to be the renormalized quasi-particle energy so that the interaction Hamiltonian contains an energy counter term, which makes the energy shift vanish:

$$\delta\omega_k(\beta) = 0. \tag{8.47}$$

Since $\delta\omega_k(\beta)$ is a function of ω_k, this equation is the quasi-particle energy equation which determines ω_k. This situation is the same as the energy renormalization in the usual quantum field theory.

The imaginary part of Σ with $k_0 = \omega_k(\beta)$ is the dissipative term of the quasi-particle. Equation (8.42) gives the following dissipative term:

$$\Im\left[\Sigma\left(k_0 = \omega_k(\beta), \vec{k} : \beta\right)\right] = -i\kappa_k(\beta)A(\omega_k(\beta)). \tag{8.48}$$

where

$$\kappa_k(\beta) = \pi\sigma_k(\omega_k(\beta)), \tag{8.49}$$

and

$$A(\omega_k) = B^{-1}(\omega_k(\beta))\tau_3 B(\omega_k(\beta)). \tag{8.50}$$

Since the tilde field carries the negative hat-energies, higher order terms in the loop expansion always make κ non vanishing whenever certain interactions exist. Thus all of the quasi-particles are unstable due to thermal dissipation [2, 3]. This makes TFD very different from the usual quantum field theory.

Adding the self-energy term to the free Hamiltonian we obtain

$$\hat{H}_{prop} = \hat{H}_0 + \hat{H}_{dis}, \tag{8.51}$$

where $\hat{H}_0$ is the free Hamiltonian

$$\hat{H}_0 = \int d^3k\,\omega_k(\beta)[a_k^\dagger a_k - \tilde{a}_k^\dagger \tilde{a}_k] \tag{8.52}$$

$$= \int d^3k\,\omega_k(\beta)\bar{a}_k a_k + \text{c-number} \tag{8.53}$$

$$= \int d^3k\,\omega_k(\beta)\bar{\xi}_k \xi_k + \text{c-number} \tag{8.54}$$

and

$$\hat{H}_{dis} = -i\int d^3k\,\kappa_k(\beta)\bar{a}_k A(\omega_k(\beta))a_k \tag{8.55}$$

$$= -i\int d^3k\,\kappa_k(\beta)\bar{\xi}_k(\beta)\tau_3\xi_k(\beta). \tag{8.56}$$

The $\hat{H}_{prop}$ is the effective Hamiltonian which controls the propagation character of the propagator $D_c(k)$ with $k_0 = \omega_k(\beta)$ [2, 3]. Note that $\hat{H}_{prop}$ is diagonal in terms of ξ-operators, indicating that the Bogoliubov transformation fully diagonalizes $\hat{H}_{prop}$.

The dissipative nature of quasi-particles raises some fundamental questions. The first question is concerned with the origin of the thermal instability of excitations. When there is a net flow of dissipative energy to an outside system, the origin is obvious. However, the consideration in this section indicates that the dissipation of excitations occurs even in an isolated system of interacting quantum fields. Physically what is happening is that if we introduce an excitation in our system, then through the interactions the system returns to equilibrium and the energy of the excitation is seemingly lost. Hence the term dissipation. Of course, the energy is not really lost. It is just that because a system of quantum fields is an infinite system with an infinite number of degrees of freedom, the system can absorb finite changes in energy without changing its equilibrium state. We can think of this as the wave packet of the excitation spreading out infinitely thin over the entire space, at the end of such a process the energy of the excitation has, from a practical point of view, disappeared from our system. This explains why an isolated system of interacting quantum fields without any form of external reservoir dissipates [4]. It is possible that during a dissipative process excitation energy is converted to a heat energy which also spreads over the space and is lost at the end.

The second question asks how the quasi-particle picture should be modified by thermal effects. Since an unstable particle has a continuous energy spectrum, Landsman formulated a theory in which the unperturbed Hamiltonian is a sum of the free Hamiltonian with continuous energy [5, 6]. The perturbative calculations of this formalism give rise to results which are similar to the one given in (8.40). We do not touch this approach. For simplicity, we proceed with the use of the free quasi-particle with energy $\omega_k(\beta)$. In ordinary quantum field theory without the thermal degree of freedom the quasi-particles are the asymptotic particles and the unperturbed Hamiltonian and the dynamical map of the Hamiltonian should be the usual free Hamiltonian of the asymptotic particles. In TFD it is obvious that the quasi-particles cannot be the asymptotic particles because the quasi-particles are thermally unstable. Indeed, it was shown in [7] that, if we assume the existence of asymptotic particles and introduce the S-matrix connecting incoming to outgoing particles, the S-matrix would be the unit matrix. This indicates that quasi-particles are not the asymptotic particles.

With our choice of nondissipative free quasi-particles, we find that the dynamical map of $\hat{H}$ is $\hat{H}_0$, when we apply the same argument as the one in chapter 3, which was based on space and time translational invariance of the vacuum. Since this invariance holds true in equilibrium TFD, we may argue that the dynamical map of $\hat{H}$ is $\hat{H}_0$. Note however, that this proof is based on our choice of nondissipative free quasi-particle with energy ω_k. In the next subsection we derive the Kubo-Martin-Schwinger relation by using this equality between $\hat{H}$ and $\hat{H}_0$.

The third question is the following: is the unperturbed Hamiltonian the free one $\hat{H}_0$ or the dissipative one $\hat{H}_{prop}$? The philosophy of this book is to use $\hat{H}_0$ for the unperturbed Hamiltonian. However, this view does not always hold true. When a dissipative effect is not perturbative, it is reasonable to try $\hat{H}_{prop}$ for the unperturbed Hamiltonian [4].

It is a sensible question to ask how the use of $\hat{H}_{prop}$ for the unperturbed Hamiltonian is consistent with the statement that the dynamical map of the Hamiltonian is $\hat{H}_0$ in which the quasi-particles are free ones. To answer this question let us recall the general form of the Bogoliubov matrix in (8.90) and show that the dissipative effect can be treated by the new Bogoliubov transformation in which we add $\kappa_k t$ to the old value of the parameter s_k. Thus we introduce the dissipative Bogoliubov matrix,

$$B_{k,dis}(t) = e^{\kappa_k t \tau_3} B_k, \qquad (8.57)$$

in which B_k is the nondissipative Bogoliubov matrix used up to now. Thus, the new Bogoliubov transformation reads as

$$a_k(t)^\mu = B_{k,dis}^{-1}(t)^{\mu\nu} \xi_k(t)^\nu, \qquad (8.58)$$

$$\bar{a}_k(t)^\mu \;=\; \bar{\xi}_k(t)^\nu B_{k,dis}(t)^{\nu\mu}, \tag{8.59}$$

where

$$\xi_k(t)^\mu \;=\; e^{-i\omega_k t}\xi_k^\mu, \tag{8.60}$$

$$\bar{\xi}_k(t)^\mu \;=\; e^{i\omega_k t}\bar{\xi}_k^\mu. \tag{8.61}$$

These relations lead to

$$[\frac{d}{dt} + i\omega_k + \kappa_k A_k]a_k(t) = 0, \tag{8.62}$$

with the matrix

$$A_k \;=\; B_{k,dis}^{-1}(t)\tau_3 B_{k,dis}(t) \tag{8.63}$$

$$=\; B_k^{-1}\tau_3 B_k. \tag{8.64}$$

Note that, although $B_{k,dis}(t)$ depends on time, A_k does not: a change of s_k does not modify the A-matrix. The above equation shows that the Hamiltonian for this field is $\hat{H}_{prop} = \hat{H}_0 + \hat{H}_{dis}$ which is the dissipative Hamiltonian. However the quasi-particle operator $\xi_k(t)$ satisfies the nondissipative equation

$$[\frac{d}{dt} + i\omega_k]\xi_k(t) = 0, \tag{8.65}$$

implying that the quasi-particle is still free. The point here is that the dissipative nature is not associated with the quasi-particle operators but with the a_k-operators: a time-dependent change of the Bogoliubov matrix modifies the equation for the a_k-operators. The same situation appears in chapter 9 where the time-dependent TFD will be discussed. This argument seems to justify the use of $k_0 = \omega_k$ for the on-shell condition used in this book. The on-shell condition for dissipative particles in TFD is a controversial subject.

Perturbation calculation of the dissipative coefficient κ_k of a charged fermion in a QED hot plasma is known to give rise to an infrared catastrophe when the free unperturbed Hamiltonian is used. This suggests that the dissipative effect in this case is not perturbative. This is in conformity with a result of Pisarski [8] which contained terms with $e^2 \ln[1/e]$ (e: the electric charge). However use of $\hat{H}_{prop}$ for the unperturbed Hamiltonian in this problem has not been tried yet.

There is an interesting study of dissipative quasi-particle, obtained by solving an exactly solvable model, the Thirring model. The result at finite temperature showed a rather peculiar distribution of complex multipole singularities in a one body propagator [9]. However, the authors cautioned that this peculiarity might be caused by the intrinsically unperturbative nature of the model at zero temperature.

Closing this subsection let us note that, when we treat a transient system with a net flow of dissipative energy to an outside system, $\hat{H}_{prop}$ is a useful choice for the unperturbed Hamiltonian.

8.1.4 PARTICLE NUMBER AND THE NUMBER PARAMETER

So far the particle number parameter has been defined by the one which appears in the thermal Bogoliubov transformation. This is the particle number defined in the interaction representation:

$$n(\omega_k)\delta(\vec{k}-\vec{l}) = \langle 0(\theta)|a_k^\dagger a_l|0(\theta)\rangle. \tag{8.66}$$

However, the observed number should be given by the expectation value of the Heisenberg number operator:

$$n_{Hk}\delta(\vec{k}-\vec{l}) \equiv \langle 0(\theta)|a_{Hk}^\dagger a_{Hl}|0(\theta)\rangle. \tag{8.67}$$

Here, the suffix H means the Heisenberg operators. This quantity is the (11)-component of the corrected one body propagator $D_c(t-t')^{\mu\nu}$ at the limit $t \to t'$. Thus n_{Hk} is obtained from the spectral representation (8.40) applied to D_c given in the last subsection. The result is

$$n_{Hk} = \int d\kappa\, \rho_k(\kappa)n(\kappa). \tag{8.68}$$

According to (8.43) the spectral function ρ_k is a function with a peak at $\kappa = \omega_k$ and width $\kappa_k(\beta)$. Thus the observed number distribution has a fluctuation of order of $\kappa_k(\beta)$ around the Boltzmann distribution n_k.

8.1.5 THE KUBO-MARTIN-SCHWINGER CONDITION

In this subsection we derive a rule for exchange of two operators in their vacuum expectation values which reads as

$$\langle 0(\theta)|A(t)B(t')|0(\theta)\rangle = \langle 0(\theta)|B(t')A(t+i\beta)|0(\theta)\rangle. \tag{8.69}$$

This relation is called the Kubo-Martin-Schwinger (KMS) condition [10, 11].

In TFD the derivation of this relation is based on the thermal state condition [12, 1]. Note that in derivation of these thermal state conditions, it was assumed that $\hat{H}$ is weakly equal to $\hat{H}_0$.

Since $\hat{H}$ generates the time translation, we can rewrite the thermal state conditions in (8.18) and (8.19) as

$$[A(t) - \tilde{A}^\dagger(t-i\alpha\beta)]|0(\theta)\rangle = 0, \tag{8.70}$$

$$\langle 0(\theta)|[A(t) - \tilde{A}^\dagger(t-i(1-\alpha)\beta)] = 0. \tag{8.71}$$

We can now proceed as follows:

$$\begin{aligned}
\langle 0(\theta)|A(t)B(t')|0(\theta)\rangle &= \langle 0(\theta)|\tilde{A}^\dagger(t+i[1-\alpha]\beta)B(t')|0(\theta)\rangle \\
&= \langle 0(\theta)|B(t')\tilde{A}^\dagger(t+i[1-\alpha]\beta)|0(\theta)\rangle \\
&= \langle 0(\theta)|B(t')A(t+i\beta)|0(\theta)\rangle. \tag{8.72}
\end{aligned}$$

This completes the derivation of the KMS condition. In statistical mechanics this relation is derived from the trace formulae for ensemble average [10, 11]. In C*-algebra approach this condition was an axiom. In TFD this condition is a result of the thermal state condition, which originates from the structure of the Hamiltonian $\hat{H}$ and the tilde conjugation rules.

8.1.6 FREE ENERGY

The TFD originated from the equality between the vacuum expectation value and the thermal average $Tr[A\rho]/Tr[\rho]$. However, the free energy F does not have this form, because it is given by

$$F = -\beta^{-1} \ln Tr[\exp[-\beta H]]. \tag{8.73}$$

Thus we need a trick to enable TFD to get at the free energy.

To calculate this in TFD we note that the dynamical Hamiltonian consists of free and interaction parts:

$$H = H_0 + H_I. \tag{8.74}$$

Now introduce

$$H(s) = H_0 + sH_I \tag{8.75}$$

and

$$F(s) = -\beta^{-1} \ln Tr[\exp[-\beta H(s)]] \tag{8.76}$$

with real c-number parameter s. We then have

$$\frac{d}{ds} F(s) = \frac{Tr[e^{-\beta H(s)} H_I]}{Tr[e^{-\beta H(s)}]}, \tag{8.77}$$

which can be calculated in TFD as

$$\frac{d}{ds} F(s) = \langle 0(\beta, s)| H_I |0(\beta, s)\rangle. \tag{8.78}$$

Here $|0(\beta, s)\rangle$ is the thermal vacuum at temperature β^{-1} associated with the Hamiltonian $H(s)$. Integrating this equation, we obtain $F = F(1)$ as

$$F = F(0) + \int_0^1 ds \langle 0(\beta, s)| H_I |0(\beta, s)\rangle. \tag{8.79}$$

This is the relation in the Heisenberg representation.

In the interaction representation, we have

$$F = F(0) + \int_0^1 ds \langle 0(\beta)| T[H_I(t_0) \hat{S}] |0(\beta)\rangle. \tag{8.80}$$

Here we made use of the notations used in the formulation of the perturbative calculation presented in the last chapter. The $|0(\beta)\rangle$ is the thermal vacuum at temperature β^{-1} in the interaction representation. This is the TFD method for calculation of the free energy F [13].

8.1.7   THE SPONTANEOUS BREAKDOWN OF SYMMETRIES

In chapter 5 we studied the mechanism of spontaneous breakdown of symmetries (creation of order) at zero temperature. In chapter 6 we studied how particle condensation in vacua creates macroscopic objects with quantum mechanical degrees of freedom. All these stories do not require any modification at finite temperature in TFD. Indeed, TFD naturally extends the zero temperature calculation to finite temperature in these problems. This aspect of TFD is frequently helpful to us in formulating analysis at finite temperature in a well organized form.

When a system has a solution in which there is some spontaneously broken symmetry with certain NG particles at zero temperature, the same symmetry breakdown at finite temperature also creates the NG particles, although their energy spectra depends on temperature. As was shown in chapter 5, the NG particles at zero momentum do not participate in any reactions. The same proof holds true at finite temperature. However, the NG particles with nonvanishing momentum are thermally unstable. Thus the NG particles are stable only at zero momentum.

The structure of symmetry rearrangement and the Ward - Takahashi relations are carried over to finite temperature cases without any modification [1], although every matrix element in these relations depends on temperature.

Therefore, we do not need any modification in our consideration of the spontaneous symmetry breakdown and creation of macroscopic objects at finite temperature. We wish to note only that with the thermal degree of freedom the story of ordered states, macroscopic objects and phase transition becomes realistic because phase transitions usually occur at finite temperature. In other words, practical study of spontaneously broken symmetries should take into account temperature effects. When we proceed along this line of thought, we are naturally led to time-dependent thermal problems. Since this is a developing subject, it will be touched on in the next chapter.

8.1.8   THE THERMAL BREAKDOWN OF SYMMETRIES

There are some symmetries which are naturally broken even by the slightest thermal effect. The most well known one is the breakdown of the Lorentz symmetry in relativistic models at finite temperature. Another example is the breakdown of supersymmetry at finite temperature.

In statistical mechanics the breakdown of Lorentz symmetry is understood to be due to the presence of a particular coordinate in which the heat bath is at rest. However, in TFD one can include the heat bath as a part of the systems, if one wishes so. Then we are dealing with an isolated system. In this case the Lorentz symmetry breakdown takes the form of a spontaneous one [14, 15].

In case of the supersymmetry the origin of symmetry breakdown is the difference between the boson and fermion number distribution; this difference breaks the boson-fermion symmetry. This too is a case of spontaneous symmetry breakdown [16, 17, 18, 19, 20, 21, 22, 23, 24].

These symmetry breakdowns are called the thermal breakdown of symmetries [15].

A characteristic feature of the thermal breakdown of symmetries is that the breakdown occurs even in cases with no interactions. On the other hand the Nambu-Goldstone theorem requires the appearance of a zero energy level in any spontaneous breakdown of symmetry. Since the Hamiltonian in TFD is not H but $\hat{H}$, the NG theorem requires, not zero energy states, but zero hat-energy states. A question to be asked is how these NG modes are created even without interaction.

To answer to this question, it is useful to recall a theorem from the usual quantum field theory without thermal degrees of freedom presented in chapter 3; it states that the dynamical map of a time independent generator of an invariant symmetry consists of terms of the form $\xi_k^\dagger \xi_k$ and terms linear in energy gapless fields. In TFD this theorem needs to be modified; since tilde particles carry negative hat-energy, terms of the form $\xi_k \tilde{\xi}_k$ and their dagger conjugate carry zero hat-energy, and therefore, these terms are permitted to appear in the dynamical map of time independent generators. In the case of the thermal breakdown of the Lorentz symmetry, these are the NG modes [15].

In case of the thermal breakdown of supersymmetric models without interactions, these terms have the form $\xi_k^F \tilde{\xi}_k^B$, $\xi_k^B \tilde{\xi}_k^F$ and their dagger conjugates. Here the superscripts B and F refer to boson and fermion respectively. Thus these terms are the zero hat-energy fermionic modes, substituting for the Goldstino [24]. When interactions appear, the supersymmetry breakdown usually creates an energy difference between the fermion and boson, making the above terms not of zero hat-energy. In such cases these fermionic pairs are replaced by real Goldstinos [20, 25].

8.1.9   THERMAL QUANTUM FIELD THEORIES

There have been several formulations of thermal quantum field theories. In this section their relation to TFD is briefly summarized. (For the subject of thermal quantum field theories in general, see the article in [26]. More recent developments were reviewed in the proceedings of the first [27] and the second [28] international workshops on "Thermal Quantum Field Theories and Their Applications".)

The C*-algebra approach:

An algebraic theory for equilibrium thermal physics has been formulated in the framework of C*-algebras [29, 30]. There a Gibbs state is given by an expectation functional $\omega(A)$ of operators which belong to the C*-algebra. In this thermal theory the KMS condition is also used as a fundamental axiom.

Equivalence between TFD with its equilibrium thermal state conditions and this C*-algebra approach was proved by Ojima in his celebrated paper [31]. It remains to be seen if a future extension to nonequilibrium situation could stay in the framework of the C*-algebra. In this book we do not address to such a question; we pursue such an extension using a TFD approach.

The Liouville equation and superoperator formalism:

The basic relations of statistical mechanics may be summarized by two relations. One is the Liouville equation

$$i\frac{\partial}{\partial t}\rho = [H, \rho] \tag{8.81}$$

and another is the well known trace formula discussed in 2.4:

$$\langle A \rangle = Tr[\rho A]/Tr[\rho]. \tag{8.82}$$

The ρ is the density matrix. The superoperator formalism was developed in order to put the Liouville equation in the form of the Schrödinger equation. To do this, the density matrix is treated as a vector in a vector space on which an operator can operate in two ways; one is operating on the left side of ρ as $A\rho$, while the other is on the right side of ρ as ρA. We may rewrite the latter as the left side operation $\tilde{A}\rho$. In this way, every degree of freedom is doubled: with an operator A we associate with it its tilde conjugate $\tilde{A}$. The doublet operators are frequently called superoperators [32, 33]. The correspondence between the superoperator formalism and TFD were clarified in [34, 2, 3, 35, 36]. The α-degree of freedom was first discovered by Schmutz in his analysis of this correspondence [34]. The above argument gives

$$[H, \rho] = (H - \tilde{H})\rho, \tag{8.83}$$

which leads to the familiar form for the TFD Hamiltonian $\hat{H}$: $\hat{H} = H - \tilde{H}$.

To make the point of our consideration clear, we assume a free oscillator $(a, a^\dagger)$ satisfying $[a, a^\dagger] = 1$. As we did in section 2.4, we put the density matrix in the form

$$\rho = f^{a^\dagger a}, \tag{8.84}$$

Let $|m\rangle$ denote the eigenvector of $a^\dagger a$ with eigenvalue m. We can then write the density matrix as

$$\rho = \sum_m |m\rangle f^m \langle m|. \tag{8.85}$$

We now have

$$[H, \rho] = \sum_m [\, H|m\rangle f^m \langle m| - |m\rangle f^m \langle m|H \,]. \tag{8.86}$$

To move to TFD we perform the replacement [35]:

$$|m\rangle\langle m| \rightarrow |m, \tilde{m}\rangle. \tag{8.87}$$

In this way the right side operation becomes a left side one.

This shows that the structure $\hat{H} = H - \tilde{H}$ of the TFD Hamiltonian is closely related to the Liouville equation in statistical mechanics. Recall that in TFD this structure came from the tilde invariant nature of the thermal vacuum.

The Path Ordered Method:

The most popular method based on real-time thermal Green's functions is the so called closed time path method or the Schwinger - Keldysch (SK) method [37, 38, 39, 40, 41, 42]. This is in contrast to TFD which is basically an operator formalism, although the propagators can be derived from it. A similarity between the perturbation scheme in TFD and the one in the SK method was first pointed out in [43, 44].

The SK method corresponds to a particular choice in the general path ordering method (POM) [45]. In POM the thermal multipoint propagator at an equilibrium situation is given by

$$\langle T_c[\psi_H(x_1) \cdots \psi_H(x_n)]\rangle = \frac{\langle \psi(x_1) \cdots \psi(x_n) \exp\left(-i \int_c dz H_I(z)\right)\rangle}{\langle T_c \exp\left(-i \int_c dz H_I(z)\right)\rangle}. \quad (8.88)$$

Here T_c is the path ordered product on a path in the complex time plane running from t_- to $t_- - i\beta$; it goes along the path

$$\begin{aligned} t_- &\rightarrow t_+ \rightarrow t_+ - i(1-\alpha)\beta \\ &\rightarrow t_- - i(1-\alpha)\beta \rightarrow t_- - i\beta. \end{aligned} \quad (8.89)$$

This path is shown in the Figure 2.

When the interaction effect from the infinitely far domains vanishes, the contributions from the vertical lines in the denominator and numerator on the right hand side of (8.88) cancel. Then the contributions come only from two lines along the real axis; one goes above the real axis while another returns a distance $(1-\alpha)\beta$ below the real axis. When we write ψ above the real axis as ψ^1 and the one below the real axis as ψ_2, we obtain the doubled components. Then the two point propagator in the POM becomes the two-by-two matrix [37, 38, 46]. This turns out to be the same as the one in TFD when we regard α above as the α parameter in TFD. Recall that the physical observables in equilibrium TFD are independent of α. Thus, in POM too, the observables are independent of α when we consider an equilibrium situation. Note that *this similarity between TFD and POM is confined only to the Green's functions* ; in TFD the tilde fields commute or anticommute with non tilde fields, while in POM the operators on the path above the real axis do not commute or anticommute with those on the path below the real axis. The basic framework of TFD is entirely different from POM.

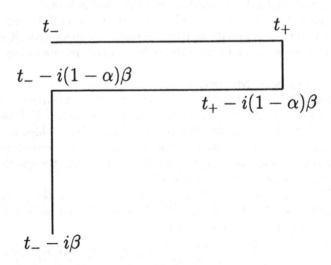

Figure 2. Picture for the path in (8.89).

Semenoff, who knew of TFD, later worked with Niemi and formulated, through use of POM, an effective potential method in their celebrated paper [47]. This paper is the first to derive $\alpha = 1/2$-POM.

In TFD we found that two choices, $\alpha = 1$ or $\alpha = 1/2$, are particularly convenient. In the SK method the choice $\alpha = 1$ has been used. Since this gives a closed path around the real axis, the SK method is called the closed path formalism (CPF). The CPF has a long history. An excellent description of the early development is given by Kadanoff and Baym in [41]. The article in [48] is also recommended to readers. Its application to nuclear physics is given in many papers; for example, see [49, 50]. Applications to the physics of plasma [51], and to astrophysics and cosmology [52] are only some examples among a vast number of publications using CPF. The close relation between the Kadanoff-Baym formalism and TFD was clarified by T. Arimitsu [53].

The formulation of dissipative quasi-particles in CPF raises a significant problem as it did in TFD, and this subject is still controversial (see, for example, [54]). CPF has been considered to be a useful method for nonequilibrium problems and a concrete model calculation by Korenman [55] is well known. A TFD treatment of the Korenman model was presented in [56].

The Imaginary Time Formalism:

All the formalisms mentioned above use real time while temperature is an additional parameter. On the other hand, in the imaginary time formalism the time with imaginary unit is linked to the temperature. This formalism was started by Matsubara [57] and in many areas of physics this is still the most popular method. The most significant aspect of this formalism in the history of thermal quantum field theories is that this formalism is the one which opened the whole subject of thermal quantum field theories. The derivation of the method was rather simple. Around 1955 a perturbative calculational formulation for the operator of time translation, $\exp[iHt]$, was being actively developed. The imaginary time formalism proposed to replace the time t by $i\beta$ in order to obtain the matrix elements of the equilibrium density matrix operator $\exp[-\beta H]$. This corresponds to the above mentioned path ordered method with $t_- = t_+ = 0$ so that the path runs along the imaginary time axis from zero to $-i\beta$. A large difference between the usual field theory and the imaginary time formalism is that in the usual quantum field theory time covers the entire domain from $-\infty$ to ∞, while in the imaginary formalism it covers the imaginary domain of only a finite segment. Therefore, it was not obvious if the Feynman expansion in the Fourier representation was usable in the imaginary time formalism. This question was answered in two articles, [58] and [59], which appeared independently and which started practical applications of the imaginary time formalism using the Feynman diagram method. The Fourier transform was found to have the form of a sum over discrete energies called the Matsubara frequencies. The Matsubara frequencies were discovered in these two articles. An interrelationship between these two articles was mentioned in a footnote in one, [58], of these two papers. An excellent description of the development of the imaginary time formalism in its early days was given by Ezawa in [60]. The derivation of the Feynman diagram method needed the rule for Wick products and complete proof of this rule in the imaginary time formalism was given in [61]. Bloch and his collaborator [62] opened a large program of development of thermal Green's function theory motivated by [59].

Imaginary time formulation has some limitation. To move from the imaginary time formalism to real time formalisms needs an analytical continuation which is sometimes tricky. Also the imaginary time formalism is a Green's function formalism which is confined to the stationary equilibrium situations.

TFD applications:

The construction of the formalism of TFD has been made by many physicists in parallel to the development of other real time formalisms. Applications of TFD have been vigorously pursued and it is impossible to cover in this book even just the important contributions. In the study of two mode squeezed states, TFD was used by many physicists as a basic technique, as was described in chapter 2 and 3. A similar application was in the subject of parametric amplifiers. For the TFD applications in these subjects, see also [63, 64, 65]. In nuclear physics the concept of temperature is significant and TFD has been used by several people. In particular, the application of TFD to the random phase approximation (RPA) for study of giant resonances, in [66] and other articles, showed that RPA in TFD carries more information than the imaginary time formalism, because each loop is not a single function but a two by two matrix containing four functions. The fact that TFD is an operator formalism is also found to be useful in nuclear physics where a variety of canonical transformations are employed. This can be seen, for example, in the recently developed TFD Hatree-Fock method [67, 68, 69]. Wherever a process in which an average over certain probability distribution is needed, TFD methods can be used. This can be seen in the development of a dynamical approach to spin glass problem [70, 71, 72, 73]. On the basis of TFD, M. Suzuki developed highly elaborate techniques in many areas of statistical physics (see, for example, [74]). These are just some examples of the applications of TFD.

There are many technical aspects associated with perturbative calculations in TFD (and in real time formalisms in general) and we can only indicate a few here. Imaginary parts and decay rates can be obtained using a method developed by Kobes and Semenoff [75, 76] and similar problems were considered by Danielewicz [77] and Niegawa [78]. The relationship between the various types of Green's functions one encounters in problems, time ordered, retarded etc., are discussed by Evans [79] and Kobes [80, 81]. Free energies or effective potentials are very useful quantities. These have been studied extensively, for example see the work of Fujimoto and collaborators [82, 83, 84].

The TFD treatment of strings and superstrings at finite temperature has also been studied by several physicists [85, 86, 87, 88, 89, 90].

For a more complete survey see the "Proceedings of the Workshops on Thermal Quantum Field Theories and their Applications" held at Cleveland, 1988 and at Tsukuba, Japan, 1990 [27, 28].

Temperature and the Microscopic World:

In nuclear physics the concept of temperature is a traditional concept. There, the argument is that when the number of nucleons is large, one can see only results after a coarse graining process, which gives rise to the temperature. However, as was explained in subsection 6.5.3, a nucleus is an isolated system which self-consistently creates a boundary surface which encloses nucleons. Therefore, to be precise, we may need a thermal theory for an isolated system. An approach along this line of thought leads us to the use of thermal quantum field theories. Traditionally, statistical mechanics approaches based on the use of the density matrix and the imaginary time formalism are the ones which have been most commonly used. Nowadays, the path ordered method and TFD are also becoming popular. Some of them were mentioned above when we discussed TFD applications. However almost all the applications use equilibrium theories, although it is obvious that the thermal situation changes in time through nuclear collision processes. It is a big challenge to formulate such a thermal theory for reactions of microscopic particles.

Once we accept the view that thermal effects should be considered in microscopic particle reactions such as nuclear reactions, the subject can be extended naturally to include high energy particle collisions. The idea of introducing temperature into high energy particle physics started around 1950. The subject at issue was the multiple production of mesons by nuclear collisions in the high energy region above 10^{11} eV, which was considered to be very large in those days. At that time a new trend of statistical theory for this phenomenon was opened by Fermi [91, 92] and Landau [93, 94]. Since many particles participate in this kind of phenomena, it is hard to describe behavior of each particle separately and some kind of statistical averaging process is required. Fermi assumed that collision of high energy nucleons would produce a hot meson-nucleon soup confined in a domain which forms a disc due to the Lorentz contraction caused by high velocity of colliding nucleons. He proposed to treat this system by a statistical method. Landau added to this a hydrodynamical stage in which the disc expands in the direction perpendicular to the disc. This expansion causes a cooling down process through which the final state particles are materialized. To solve this problem one needs an equation of state for the meson nucleon soup. With the assumption that this soup is an extremely hot ideal gas the equation $p = \epsilon/3$ was made use of. Here p and ϵ are the pressure and energy density respectively. To study the interaction effects in this equation of state the authors of the article [59] applied the imaginary time formalism to an interacting quantum field model. This may have been the first work in which thermal quantum field theories were applied to high energy particle reactions. Since it is very difficult to describe the time dependent thermal changes through this complex process, it was assumed that the situation can be approximated crudely by an equilibrium state with a temperature. This study exposed several aspects of thermal quantum field theories. First, the Fourier representation of the Feynman diagram method is usable when the energy integration is replaced by the sum over the discrete frequencies which are now called the Matsubara frequencies. It also showed that, when a model is renormalizable at zero temperature in the sense that all of the ultraviolet divergencies can be eliminated by the renormalization method, it is renormalizable also at finite temperature (in this book this was explained at the end of subsection 7.2.6).

This idea has now been inherited by QCD physics, in which it is thought that at high temperature and high pressure hadronic matter undergoes a phase transition to a hot gas plasma of free gluons and quarks. Here too, many of the calculations have been confined to an equilibrium state [95, 96, 97, 98, 99, 100]. This is an active subject and it is not tried here to cover most of the references. See for example the proceedings of the "Quark Matter Conferences" [101, 102]. Note that this subject is also concerned with thermal phenomena of particle reactions taking place in an isolated system, and also that a precise description requires a time dependent thermal theory.

A similar line of approach has been applied also to a very different subject, that is the study of evolution of the early Universe. This subject also is concerned with time dependent thermal process in an isolated system. For references see for example the proceedings of Workshops on " Thermal Quantum Field Theories and their applications" [27, 28] mentioned previously.

Temperature and Renormalization Group:

The renormalization group method [103, 104] is one of the powerful methods used to analyze the asymptotic behavior of physical quantities, and has been applied to many areas of physics. The most successful applications may be found in the analysis of deep inelastic scattering (see, for example, [105]) and in the analysis of critical behavior in phase transitions [106].

The traditional form of the renormalization group method scales the energy-momentum; temperature effects were treated rather phenomenologically. However it is logically more consistent to scale both the temperature and energy-momentum separately, because then one can deal with both dynamical and temperature dependent behavior. This kind of formalism is called the temperature renormalization group method, which deals with, not a one, but a two parameter Abelian group. For detailed accounts of the formulation and applications, see, for example, [107, 108, 109, 110, 111, 112, 113, 114].

8.2   The t-Representation

8.2.1   INTRODUCTION

The story given in this chapter up to this point is the commonly accepted form of equilibrium TFD. However, one of our questions is to ask what kind of thermal state quasi-particles go through in a short time. In quantum theory all kinds of quantum states exist as a fluctuation for a short time. Therefore, it is natural to expect that in TFD this fluctuation includes thermal states. If we were to answer the above question for two particle states, we would know if it is possible to have a quark-gluon plasma in high energy heavy ions collision or not. In this section we study the above question for one particle. The Fourier representation employed up to now is not convenient for this purpose; since we want to follow the behavior of a quasi particle in time, we need to work in the time representation rather than in its Fourier conjugate. We call this the t-representation [115].

8.2.2   PREPARATION

As a preparation for the t-representation we introduce some useful relations in this subsection [115].

According to (7.86) the general form of the Bogoliubov matrix is

$$B_k(\theta)^{\mu\nu} = (1 + \sigma n_k)^{1/2} e^{s_k \tau_3} \begin{bmatrix} 1 & -f_k^\alpha \\ -\sigma f_k^{1-\alpha} & 1 \end{bmatrix}, \qquad (8.90)$$

where σ is plus for boson and minus for fermion. The f_k is

$$f_k = \frac{n_k}{1 + \sigma n_k}. \qquad (8.91)$$

which gives $n_k = f_k/[1 - \sigma f_k]$. We choose α_k to be independent of $\vec{k}$ and denote it by α. The inverse of the above matrix is

$$B_k^{-1}(\theta)^{\mu\nu} = (1 + \sigma n_k)^{1/2} \begin{bmatrix} 1 & f_k^\alpha \\ \sigma f_k^{1-\alpha} & 1 \end{bmatrix} e^{-s_k \tau_3}, \qquad (8.92)$$

In order to apply the Feynman diagram method we choose $\alpha = 1$. As was shown in section 8.1.1 the form of the Bogoliubov matrix is further simplified by choosing the parameter s_k to be

$$s_k = \ln[1 + \sigma n_k]^{1/2}, \qquad (8.93)$$

which gives

$$e^{s_k \tau_3} = \begin{bmatrix} [1 + \sigma n_k]^{1/2} & 0 \\ 0 & [1 + \sigma n_k]^{-1/2} \end{bmatrix}. \qquad (8.94)$$

With this, the relation (8.90) for the Bogoliubov matrix takes the remarkably simple form

$$B_k = \begin{bmatrix} 1 + \sigma n_k & -n_k \\ -\sigma & 1 \end{bmatrix}, \qquad (8.95)$$

the inverse of which is

$$B_k^{-1} = \begin{bmatrix} 1 & n_k \\ \sigma & 1 + \sigma n_k \end{bmatrix}. \qquad (8.96)$$

Now a straightforward matrix calculation gives

$$\frac{1 + \tau_3}{2} B_k(t) = \frac{1 + \tau_3}{2} + \sigma n_k(t)(T_0 + T_-), \qquad (8.97)$$

$$B_k^{-1}(t)\frac{1 + \tau_3}{2} = T_+, \qquad (8.98)$$

$$\frac{1 - \tau_3}{2} B_k(t) = T_-, \qquad (8.99)$$

$$B_k^{-1}(t)\frac{1 - \tau_3}{2} = \frac{1 - \tau_3}{2} - \sigma n_k(t)(T_0 - T_+). \qquad (8.100)$$

Here the notation used is

$$T_0 = \begin{bmatrix} 1 & -\sigma \\ \sigma & -1 \end{bmatrix}, \qquad (8.101)$$

$$T_+ = \begin{bmatrix} 1 & 0 \\ \sigma & 0 \end{bmatrix}, \tag{8.102}$$

$$T_- = \begin{bmatrix} 0 & 0 \\ -\sigma & 1 \end{bmatrix}. \tag{8.103}$$

The three matrices, $T_\pm$ and T_0, play a fundamental roles in practical calculations. A remarkable fact is that these matrices satisfy some simple algebraic relations. They are

$$
\begin{aligned}
T_0^2 &= 0, &(8.104)\\
T_+^2 &= T_+, &(8.105)\\
T_-^2 &= T_-, &(8.106)\\
T_+ T_0 &= T_0 T_- = T_0, &(8.107)\\
T_0 T_+ &= T_- T_0 = T_+ T_- = T_- T_+ = 0, &(8.108)\\
T_+ + T_- &= 1. &(8.109)
\end{aligned}
$$

Using these relations we can easily show that

$$B^{-1}(n^1)\frac{1+\tau_3}{2}B(n^2) = T_+ + \sigma n^2 T_0, \tag{8.110}$$

$$B^{-1}(n^1)\frac{1-\tau_3}{2}B(n^2) = T_- - \sigma n^1 T_0, \tag{8.111}$$

$$A(n) \equiv B^{-1}(n)\tau_3 B(n) = T_+ - T_- + 2nT_0. \tag{8.112}$$

Here $B(n)$ is the Bogoliubov matrix with the number parameter n. Use of the above formulae leads to the following relation:

$$
\begin{aligned}
B^{-1}(N^1)\begin{bmatrix} g_1 & 0 \\ 0 & g_2 \end{bmatrix} B(N^2) &= (g_1 T_+ + g_2 T_-) \\
&+ \sigma(g_1 N_k^2 - g_2 N_k^1)T_0. &(8.113)
\end{aligned}
$$

The formulae in this subsection are useful in the calculation of propagators.

8.2.3 ONE BODY PROPAGATOR

We have seen in a previous section in this chapter that in equilibrium TFD one body propagators with corrections due to interactions have the spectral representation given in (8.40). Writing a one body propagator for the Heisenberg oscillator operators $a_{kH}(t)^\mu$ and $\bar{a}_{Hk}(t)^\mu$ as

$$D_k(t-t')^{\mu\nu} = \int \frac{dk_0}{2\pi} e^{-ik_0(t-t')} D_k(k_0)^{\mu\nu}, \tag{8.114}$$

we have the spectral representation

$$D_k(k_0)^{\mu\nu} = \int_{-\infty}^{\infty} d\kappa \left[B^{-1}[n(\kappa)]\frac{\rho(\kappa, \vec{k})}{k_0 - \kappa + i\epsilon\tau_3} B[n(\kappa)] \right]^{\mu\nu}, \tag{8.115}$$

where $n(\kappa)$ is given by the Boltzmann distribution with the temperature β^{-1},

$$n(\kappa) = \frac{1}{e^{\beta\kappa} - 1}. \tag{8.116}$$

Note that this $n(\kappa)$ with $\kappa = \omega_k$ is the number density parameterizing the B-matrix in the unperturbed propagator. The function ρ is the spectral function. The equal time oscillator commutation relation for $a_{kH}(t)^\mu$ and $\bar{a}_{Hk}(t)^\mu$ gives

$$\int d\kappa\, \rho(\kappa, \vec{k}) = 1. \tag{8.117}$$

Equation (8.114) with (8.115) is rewritten after the integration over k_0 as

$$D_k(t - t')^{\mu\nu} = \int d\kappa\, \left[B^{-1}[n(\kappa)]\right.$$

$$\times \left.\begin{bmatrix} -i\theta(t - t') & 0 \\ 0 & i\theta(t' - t) \end{bmatrix} e^{-i\kappa(t-t')}\rho(\kappa, \vec{k})B[n(\kappa)]\right]^{\mu\nu}. \tag{8.118}$$

We expand this in the form of (8.113) and then perform the κ-integration. The result can be reassembled in the form of a Bogoliubov transformation as [115]

$$D_k(t - t')^{\mu\nu} = B^{-1}[N_k(t - t')]^{\mu\mu'}$$

$$\times \int d\kappa\, e^{-i\kappa(t-t')}\rho(\kappa, \vec{k})\begin{bmatrix} -i\theta(t - t') & 0 \\ 0 & i\theta(t' - t) \end{bmatrix}^{\mu'\nu'} B[N_k(t - t')]^{\nu'\nu}. \tag{8.119}$$

Here

$$N_k(t - t') = \frac{\int d\kappa\, e^{-i\kappa(t-t')}\rho(\kappa, \vec{k})n(\kappa)}{\int d\kappa\, e^{-i\kappa(t-t')}\rho(\kappa, \vec{k})} \tag{8.120}$$

$$= N_{R,k} + \nu_k(t - t') \tag{8.121}$$

with

$$N_{R,k} \equiv N_k(0) = \int d\kappa\, \rho(\kappa, \vec{k})n(\kappa). \tag{8.122}$$

This defines $\nu_k(t - t')$ as

$$\nu_k(t - t') = N_k(t - t') - N_{R,k}. \tag{8.123}$$

Note the property of

$$\nu_k(t - t' = 0) = 0. \tag{8.124}$$

A significant result can be seen when we recall the Heisenberg number density n_{Hk} defined by

$$n_{Hk}\delta(\vec{k} - \vec{l}) = \langle 0|a^\dagger_{Hk}a_{Hl}|0\rangle. \tag{8.125}$$

By considering the limit $t \to t'$ with $t \geq t'$ or $t' \geq t$, we find that

$$N_{R,k} = n_{H,k}, \qquad (8.126)$$

where (8.117) has been used. This shows that $N_{R,k}$ is the correct number density. The deviation, $\nu_k(t-t')$, of corrected Bogoliubov number parameter $N_k(t-t')$ from this real number density is called the number fluctuation.

One physical implication of the above results in the equilibrium TFD can be easily seen [115]. This comes from (8.122), which provides us with the expression of the Heisenberg number density in terms of the unperturbed one. To investigate this matter, we recall the spectral representation for the self-energy in (8.42) in subsection 8.1.3:

$$\Sigma(k_0, \vec{k})^{\mu\nu} = \int d\kappa\, \sigma(\kappa, \vec{k}) \left[B^{-1}[n(\kappa)] \frac{1}{k_0 - \kappa + i\epsilon\tau_3} B[n(\kappa)] \right]^{\mu\nu}. \qquad (8.127)$$

As was argued there, the real and imaginary parts of Σ on shell, $k_0 = \omega_k$, represent the energy shift and the dissipation, respectively,

$$\delta\omega_k = \Re\left[\Sigma\left(k_0 = \omega_k, \vec{k}\right) \right]$$

$$= r(\kappa = \omega_k, \vec{k}), \qquad (8.128)$$

$$\Im\left[\Sigma\left(k_0 = \omega_k, \vec{k}\right) \right] = -i\kappa_k B^{-1}[n(\omega_k)]\tau_3 B[n(\omega_k)], \qquad (8.129)$$

where

$$r(\kappa, \vec{k}) = \int d\kappa'\, \mathcal{P} \frac{\sigma(\kappa', \vec{k})}{\kappa - \kappa'} \qquad (8.130)$$

and the κ_k is called the dissipative coefficient given by

$$\kappa_k = \pi\sigma(\omega_k, \vec{k}). \qquad (8.131)$$

We choose the unperturbed energy ω_k to be the renormalized one so that $\delta\omega_k = 0$.

The spectral function ρ is related to the spectral function σ through

$$\rho(\kappa, \vec{k}) = \frac{\sigma(\kappa, \vec{k})}{(w - \omega_k)^2 + \pi^2\sigma^2(\kappa, \vec{k})} \qquad (8.132)$$

$$\simeq \frac{1}{\pi} \frac{\kappa_k}{\{w - \omega_k\}^2 + \kappa_k^2}. \qquad (8.133)$$

The last approximate expression is obtained by replacing $\sigma(\kappa, \vec{k})$ and $r(\kappa, \vec{k})$ with their on-shell values. When κ_k is very small, ρ has a sharp peak around $\kappa = \omega_k$. Then (8.122) together with the relation (8.126) shows that $n_{H,k}$ is well approximated by the unperturbed number $n(\omega_k)$. As thermal effects become more dominant, meaning larger value of κ_k, $n_{H,k}$ deviates more from $n(\omega_k)$. This deviation is attributed to the increase in the energy uncertainty caused by thermal instability. Thus the interaction modifies $n(\omega_k)$ into $n_{H,k}$ through thermal as well as quantum fluctuations.

The results of this section [115] show that, even in equilibrium situations, a quasi-particle feels time dependent thermal fluctuation, $\nu_k(t - t')$, during its propagation.

In a similar way two particle states may feel time dependent thermal fluctuations in the t-representation. It is an interesting question if this time dependent fluctuation has anything to do with short time thermal effects in high energy particle reactions.

The t-representation formalism provides us with a useful picture for time-space dependent thermal processes. This is the subject in the next chapter.

8.3   REFERENCES

[1] H. Umezawa, H. Matsumoto and M. Tachiki, *Thermo Field Dynamics and Condensed States*, (North-Holland, Amsterdam, 1982).

[2] T. Arimitsu and H. Umezawa, *Prog. of Theor. Phys.* **74** (1985) 429.

[3] T. Arimitsu and H. Umezawa, *Prog. of Theor. Phys.* **77** (1987) 53.

[4] I. Hardman and H. Umezawa, *Annals of Phys.* **203** (1990) 173.

[5] N. P. Landsman, *Phys. Rev. Lett* **60** (1988) 1909.

[6] N. P. Landsman, *Ann. Phys.* **186** (1989) 141.

[7] H. Narnhofer, M. Requardt and W. Thirring, *Commun. Math. Phys.* **92** (1983) 247.

[8] R. D. Pisarski, *Nucl. Phys.* **B309** (1988) 476.

[9] N. Ashida, A. Niegawa, H. Nakkagawa and H. Yokota, *Physics Lett.* **B236** (1990) 450.

[10] R. Kubo, *J. Phys. Soc. Japan* **12** (1957) 570.

[11] P. Martin and J. Schwinger, *Phys. Rev.* **115** (1959) 1342.

[12] Y. Takahashi and H. Umezawa, *Collect. Phenom.* **2** (1975) 55.

[13] H. Matsumoto, Y. Nakano and H. Umezawa, *Phys. Rev.* **D31** (1985) 1495.

[14] I. Ojima, In K. Kikkawa, N. Nakanishi and H. Hariai, editors, *Proceeding of the Conference on Gauge Theory and Gravitation at Nara Japan*, Lecture Notes in Physics 176, (Springer, Berlin, 1983).

[15] H. Matsumoto, H. Umezawa, N. Yamamoto and N. J. Papastamatiou, *Phys. Rev.* **D34** (1986) 3217.

[16] A. Das, In Kowarsky, N. P. Landsman and Ch.G van Weert, editors, *Workshops on Thermal Quantum Field Theories, Physica* **158A** (1988) 1.

[17] A. Das and M. Kaku, *Phys. Rev.* **D18** (1978) 4540.

[18] K. Teshima, *Phys. Lett.* **123B** (1983) 226.

[19] D. Boyanovsky, *Phys. Rev.* **D29** (1984) 743.

[20] H. Aoyama and D. Boyanovsky, *Phys. Rev.* **D30** (1985) 467.

[21] S. Midorikawa, *Prog. Theor. Phys.* **73** (1985) 1245.

[22] L. Girardello, M. Grisaru and P. Salamonson, *Nucl. Phys.* **B178** (1981) 331.

[23] L. Van Hove, *Nucl. Phys.* **B207** (1982) 15.

[24] K. Matsumoto, M. Nakahara, Y. Nakano and H. Umezawa, *Physica* **15D** (1985) 163.

[25] Y. Leblanc and H. Umezawa, *Phys. Rev.* **D33** (1986) 2288.

[26] N. P. Landsman and Ch. G. van Weert, *Phys. Reports* **145** (1987) 141.

[27] K. L. Kowalsky, N. P. Landsman and Ch. G. van Weert, editors, *Workshop on Thermal Quantum Field Theories and Their Applications, Physica* **158A** (1988) 1.

[28] T. Arimitsu, H. Ezawa and Hashimoto, editors, *Workshop on Thermal Quantum Field Theories and Their Applications*, (North-Holland, Amsterdam, 1991).

[29] H. Araki and E. J. Woods, *J. Math. Phys.* **4** (1963) 637.

[30] R. Haag, N. W. Hugenholtz and M. Winnink, *Comm. Math. Phys.* **5** (1967) 215.

[31] I. Ojima, *Ann. Phys.* **137** (1981) 1.

[32] U. Fano, *Rev. Mod. Phys.* **29** (1957) 74.

[33] J.A. Crawford, *Nuovo Cim.* **10** (1958) 698.

[34] Z. Schmutz, *Z. Phys.* **B30** (1978) 97.

[35] S.Chaturvedi and V. Srinivasan, Operator methods for master equations in quantum optics, *Preprint* (1989).

[36] P.A.Henning, Ch. Becker, A. Lang and U. Winkler, *Phys. Lett.* **B217** (1989) 211.

[37] J. Schwinger, *J. Math. Phys.* **2** (1961) 407.

[38] L. V. Keldysh, *Sov. Phys. JETP* **20** (1965) 1018.

[39] R. A. Craig, *J. Math. Phys.* **9** (1968) 605.

[40] S. Fujita, *Physica* **30** (1964) 848.

[41] L. P. Kadanoff and G. Baym, *Quantum Statistical Mechanics*, (Benjamin, New York, 1962).

[42] K. Chou, Z. Su, B. Hao and L. Yu, *Phys. Reports* **118** (1985) 1.

[43] T. Arimitsu, J. Pradko and H. Umezawa, *Physica* **135A** (1986) 487.

[44] M. Marinaro, *Physics Reports* **137** (1986) 81.

[45] R. Mills, *Propagators for Many-Particle Systems*, (Gordon and Breach Science Publisher, New York, 1969).

[46] B. Bezzerides and D. F. DuBois, *Phys. Rev.* **168** (1968) 233.

[47] A. J. Niemi and G. Semenoff, *Ann. Phys.* **152** (1984) 105.

[48] D. F. DuBois, In W. E. Britten, editor, *Vol IXC in Lecture in Theoretical Physics*, (Gordon and Breach, New York, 1967) p. 469.

[49] W. Botermans and R. Malfliet, *Phys. Lett.* **215B** (1988) 617.

[50] M. Tohyama, *Phys. Rev.* **C86** (1983) 187.

[51] B. Bezzerides and D. F. DuBois, *Ann. Phys.* **70** (1972) 10.

[52] E. Calzetta and B. L. Hu, *Phys. Rev* **D37** (1988) 2878.

[53] T. Arimitsu, *Physica* **148A** (1988) 427.

[54] I. D. Lawrie, *J. Phys.* **A21** (1989) L823.

[55] V. Korenman, *Ann. Phys.* **39** (1969) 1387.

[56] I. Hardman, H. Umezawa and Y. Yamanaka, *Physica* **156A** (1989) 853.

[57] T. Matsubara, *Prog. Theor. Phys.* **14** (1955) 351.

[58] A. A. Abrikosov, L. P. Gorkov and I. E. Dzyaloshinskii, *Sov. Phys. JETP* **9** (1959) 636.

[59] H.Umezawa, Y.Tomozawa and H. Ezawa, *Nuovo Cimento* **5** (1957) 810.

[60] H. Ezawa, In F. Mancini, editor, *Quantum Field Theory*, (North-Holland, Amsterdam, 1985).

[61] T. Thouless, *Phys. Rev.* **107** (1957) 1162.

[62] C. Bloch and C. Dominicis, *Nucl. Phys.* **7** (1958) 459.

[63] T. Garavaglia, *Phys. Rev.* **A38** (1988) 4365.

[64] T. Garavaglia, The characteristic functions for the squeezed coherent chaotic phoyon state with application to the jaynes-cummings model, *Preprint* (1989).

[65] T. Garavaglia, *Phys. Lett.* bf 131A (1988) 151.

[66] K. Tanabe, *Phys. Rev.* **C37** (1988) 2802.

[67] M. Yamamura, J. da Providência, A. Kuriyama and C. Fiolhais, *Prog. Theor. Phys.* **81** (1989) 1198.

[68] M. Yamamura, J. da Providência, A. Kuriyama and C. Fiolhais, *Prog. Theor. Phys.* **83** (1989) 749.

[69] M. Yamamura, J. da Providência and A. Kuriyama, *Nuclear Physics* **514A** (1990) 461.

[70] T. K. Kopec, *Japanese Journal of Applied Physics*, Supplement **26** (1987) 791.

[71] T. K. Kopec, K. D. Usadel and G. Büttner, *Phys. Rev.* **B39** (1989) 12418.

[72] T. K. Kopec and P. Wrobel, *J. Phys. (London)* **2** (1990) 397.

[73] T. K. Kopec, B. Tadic, R. Pirc and R. Blinc, *Z. Phys.* **B78** (1990) 493.

[74] M. Suzuki, In H. Ezawa and S. Kamefuchi, editors, *Progress in Quantum Field Theory*, (North-Holland, Amsterdam, 1985) p. 203.

[75] R. L. Kobes and G. Semenoff, *Nucl. Phys.* **B260** (1985) 714.

[76] R. L. Kobes and G. Semenoff, *Nucl. Phys.* **B272** (1986) 329.

[77] P. Danielewicz, *Ann. Phys.* **197** (1990) 154.

[78] A. Niegawa, *Phys. Lett.* **247B** (1990) 351.

[79] T. S. Evans, *Phys. Lett.* **252B** (1991) 108.

[80] R. Kobes, *Phys. Rev.* **D42** (1990) 562.

[81] R. Kobes, *Phys. Rev.* **D43** (1991) 1269.

[82] Y. Fujimoto, R. Grigjanis and R. Kobes,   *Prog. Theor. Phys.* **73** (1985) 434.

[83] Y. Fujimoto and R. Grigjanis, *Prog. Theor. Phys.* **74** (1985) 1105.

[84] Y. Fujimoto, *Z. Phys.* **C30** (1986) 99.

[85] Y. Leblanc, *Phys. Rev.* **D36** (1987) 1780.

[86] Y. Leblanc, *Phys. Rev.* **D37** (1988) 1547.

[87] Y. Leblanc, *Phys. Rev.* **D39** (1989) 1134.

[88] E. Ahmed, *Phys. Rev. Lett.* **60** (1988) 684.

[89] H. Fujisaki, *Prog. Theor. Phys.* **81** (1989) 473.

[90] H. Fujisaki and K. Nakagawa, *Prog. Theor. Phys.* **83** (1990) 18.

[91] E. Fermi, *Prog. Theor. Phys.* **5** (1950) 570.

[92] E. Fermi, *Phys. Rev.* **81** (1951) 683.

[93] L. D. Landau, *Uspekhi Fiz. Nauk.* **56** (1955) 309.

[94] L. D. Landau, *Izv. Akad. Nauk SSSR* **17** (1953) 51.

[95] H. A. Weldon, *Phys. Rev.* **D42** (1990) 2384.

[96] E. Braaten, R. D. Pisarski and T. C. Yuan,   *Phys. Rev. Lett.* **64** (1990) 2242.

[97] E. Braaten and R. D. Pisarki, *Phys. Rev.* **D42** (1990) 2156.

[98] R. Kobes, G. Kunstatter and A. Rebhon, *Nuc. Phys.* **B355** (1991) 1.

[99] R. Pisarski, *Phys. Rev. Lett.* **63** (1989) 1129.

[100] R. Kobes and G. Kunstatter, *Phys. Rev. Lett.* **61** (1988) 392.

[101] G. Baym, P. Braun-Munzinger and S. Nagamiya, editors, *Proceedings of Qaurk Matter '88, Nucl. Phys.* **A498** (1989).

[102] J. P. Blaizot, C. Gershel, B. Pire and A. Romana, editors, *Nucl. Phys.* **A498** (1989). *Proceedings of Qaurk Matter '90, Nucl. Phys.* **A525** (1991).

[103] E. C. G. Stückelberg and A. Peterman, *Helv. Phys. Acta* **26** (1953) 499.

[104] M. Gellman and F. E. Low, *Phys. Rev.* **95** (1954) 1300.

[105] D. J. Gross, In R. Balian and J. Jinn-Justin, editors, *Method in Field Theory*, (North-Holland, Amsterdam, 1976).

[106] K. G. Wilson and J. Kogut, *Phys. Reports* **12C** (1974) 75.

[107] H. Matsumoto, Y. Nakano and H. Umezawa, *Phys. Rev.* **D29** (1984) 1116.

[108] Y. Fujimoto and H. Yamada, *Phys. Lett.* **212B** (1988) 77.

[109] H. Yamada, *Phys. Lett.* **223B** (1989) 229.

[110] H. Nakagawa, A. Niegawa and H. Yokota, *Phys. Lett.* **244B** (1990) 58.

[111] R. Baier, B. Pire and D. Schiff, *Phys. Lett.* **238B** (1990) 367.

[112] Y. Fujimoto, K. Ideura, Y. Nakano and H. Yoneyama, *Phys. Lett.* **167B** (1986) 406.

[113] K. Funakubo and M. Sakamoto, *Prog. Theor. Phys.* **76** (1986) 490.

[114] K. Funakubo and M. Sakamoto, *Phys. Lett.* **186B** (1987) 205.

[115] Y. Yamanaka, H. Umezawa, K. Nakamura and T. Arimitsu, Thermo Field Dynamics in *t*-representation, *Univ. of Alberta Preprint* (1992).

9

Time-Space Dependent TFD

9.1 Preparation

9.1.1 A MOTIVATION

In the last chapter we developed equilibrium TFD on basis of the TFD structure presented in chapter 7. The most significant and general rules in TFD are the tilde conjugation rules and the structure of the Hamiltonian written as $\hat{H} = H - \tilde{H}$.

It is obvious that the stationary situations are only particular cases of the thermal situation. However, once a thermal quantum field theory for stationary situations is established, we can compare different thermal situations. In this way we see that the thermal degree of freedom is associated with quantum field systems, because the thermal degree of freedom can become explicit only when we observe different thermal situations. It is then natural to ask how this thermal degree of freedom behaves in time-space dependent situations. To extend stationary TFD to time-space dependent TFD is what is attempted in this chapter [1, 2, 3]. There are many reasons which motivate us to do this.

The study of the t-representation in the last chapter indicated that there appear thermal fluctuations during propagation of particle waves. This suggests that many thermal states in TFD may show up during particle reactions. We then ask how these intermediate thermal fluctuations behave in *time-dependent* thermal phenomena in the t-representation. This motivates us to extend the t-representation formalism to time-dependent thermal situations.

When we have a state in which the spatial distribution of temperature is not homogeneous, it will naturally initiate a temporal change in the thermal state, trying to achieve a homogeneous situation. This is the commonly known phenomenon of heat conduction. Since this requires a thermal quantum field theory with spatially inhomogeneous vacuum, it demands an extension of thermal quantum field theories to deal with spatially inhomogeneous situations [4]. This kind of formalism for space-time-dependent thermal physics will find many applications in condensed matter physics.

Time-dependent thermal quantum theories are also useful in the study of the evolution of the Universe. Since this evolution is supposed to be controlled by all of the fundamental quantum fields and since the evolution is supposed to involve thermal effects, the need for a time-dependent thermal quantum field theory is obvious. Note that this evolution is occurring in an isolated system.

This chapter is different from the others in the sense that the ideas presented in this chapter are only the start of a new approach rather than well established material. It is hoped that the material here will stimulate the interest of readers and motivate them to investigate all kinds of new developments in this area.

One basic question raised by time-space dependent thermal problems is how a time-space dependent process is initiated. To study this question it is useful to recall how the spontaneous breakdown of a symmetry is initiated. As was shown in chapter 5, in the perturbation calculation scheme, an analysis of spontaneous breakdown of symmetry begins with an unperturbed Hamiltonian of the form $H_0 + \delta H$ in which H_0 is a symmetric free Hamiltonian and δH breaks the symmetry under consideration. In a similar manner an analysis of a time-dependent process begins with an unperturbed Hamiltonian of the form $\hat{H}_0 - \hat{Q}$, in which $\hat{Q}$ is the operator which initiates the time-change. To identify the term $\hat{Q}$ is one of important tasks in this chapter.

Since time-space dependent TFD has a complex structure, in the next section we present just time-dependent TFD, which will be extended to time-space dependent TFD in section 3. These formulations take advantage of the t-representation.

9.2   time-dependent TFD

9.2.1   TIME-DEPENDENT BOGOLIUBOV TRANSFORMATIONS

In this section we consider time-dependent thermal situations; spatial change will be studied in the next section.

In the last two chapters we presented the fundamental framework of TFD. In stationary TFD we used the thermal Bogoliubov transformation $a_k = B_k^{-1}(\theta)\xi$ for a free oscillator field a_k^μ. It is then natural to make the assumption that a time-dependent process is described by a time-dependent Bogoliubov matrix. This assumption is tried in this chapter. Thus, the Bogoliubov transformation reads as $a_k(t) = B_k^{-1}(t)\xi_k$ [5, 6]. Temporal change in thermal situations is now fully taken care of by the temporal change of oscillator operators $a_k(t)$, and the Fock space built on the time-independent vacuum associated with ξ_k^μ is independent of time. The vacuum associated with ξ_k^μ will be denoted by $|0\rangle$. Since there are uncountable choices for this vacuum, it will be self-consistently determined, as we will show later. To this end we will use a self-consistent renormalization condition. If there were no self-consistent choice for this vacuum, our assumption that a time-independent vacuum takes care of a time-dependent thermal process would fail. It turns out that there is indeed a self-consistent choice for the time-independent vacuum. From the self-consistent renormalization condition follows the Boltzmann equation.

To perform practical calculations, we move from the Heisenberg representation to the interaction representation to make use of a free quasi-particle picture by following the formulation of the interaction representation in chapter 7. The coincidence time of the Heisenberg and interaction representations is denoted by t_0. Since we want to make use of the Feynman diagram method, we cannot choose a finite t_0 according to the consideration in chapter 7. Thus we find either $t_0 = \infty$ or $t_0 = -\infty$. As in chapter 7 we have either $(t_0 = -\infty, \alpha = 1)$ or $(t_0 = \infty, \alpha = 0)$. Let us begin with the choice $t_0 = -\infty$, which is commonly used in the usual quantum field theory. We thus try choice $(t_0 = -\infty, \alpha = 1)$, hoping that this would lead us to a self-consistent choice of the time-independent vacuum.

Now the transformations in (7.63) and (7.64) become

$$\xi_k^\mu = B_k(t)^{\mu\nu} \exp\left[i \int_{t_0}^t ds\, \omega_k(s)\right] a_k(t)^\nu, \tag{9.1}$$

$$\bar{\xi}_k^\mu = \bar{a}_k(t)^\nu \exp\left[-i \int_{t_0}^t ds\, \omega_k(s)\right] B_k^{-1}(t)^{\nu\mu}, \tag{9.2}$$

where the definition of the thermal doublets is the same as the one given in the last chapter.

This requires some explanation. In a stationary thermal situation we have $a_k(t) = a_k \exp[-i\omega_k t]$. In a time-dependent situation the quantum energy ω_k depends on time due to thermal effects. Thus we introduce $\omega_k(t)$, and the wave function $\exp[-i\omega_k t]$ is replaced by $\exp[-i\int_{t_0}^t ds\, \omega_k(s)]$, because this satisfies

$$i\frac{d}{dt} e^{[-i\int_{t_0}^t ds\, \omega_k(s)]} = \omega_k(t) e^{[-i\int_{t_0}^t ds\, \omega_k(s)]}. \tag{9.3}$$

We now apply this formalism to a free field. For simplicity we consider the case:

$$\varphi(x)^\mu = \int \frac{d^3k}{(2\pi)^{(3/2)}} e^{i\vec{k}\cdot\vec{x}} a_k(t)^\mu, \tag{9.4}$$

$$\bar{\varphi}(x)^\mu = \int \frac{d^3k}{(2\pi)^{(3/2)}} e^{-i\vec{k}\cdot\vec{x}} \bar{a}_k(t)^\mu \tag{9.5}$$

with the equal time commutation relation

$$[a_k(t)^\mu, a_l(t)^\nu] = \delta_{\mu\nu}\delta(\vec{k} - \vec{l}). \tag{9.6}$$

In the usual quantum field theory the quasi-particle operators have the form $\xi_k(t) = \xi_k \exp[-i\omega_k t]$. Importing this aspect to TFD, we introduce the quasi-particle operators

$$\xi_k(t)^\mu = \xi_k^\mu \exp\left[-i \int_{t_0}^t ds\, \omega_k(s)\right], \quad \bar{\xi}_k(t)^\mu = \bar{\xi}_k^\mu \exp\left[i \int_{t_0}^t ds\, \omega_k(s)\right]. \tag{9.7}$$

Since ξ_k^μ and $\bar{\xi}_k^\mu$ are independent of time, we have

$$[\frac{d}{dt} + i\omega_k(t)]\xi_k(t)^\mu = 0, \tag{9.8}$$

$$\bar{\xi}_k(t)^\mu[\overleftarrow{\frac{d}{dt}} - i\omega_k(t)] = 0. \tag{9.9}$$

The Hamiltonian for these equations is $\hat{H}_0$. In other words the unperturbed Hamiltonian for quasi-particle is $\hat{H}_0$ which is diagonal in ξ.

The above Bogoliubov transformation now reads as

$$\xi_k(t)^\mu = B_k(t)^{\mu\nu}a_k(t)^\nu, \quad \bar{\xi}_k(t)^\mu = \bar{a}_k(t)^\nu B_k^{-1}(t)^{\nu\mu}. \tag{9.10}$$

Thus the equations (9.8) and (9.9) give [7]

$$[\frac{d}{dt} + i\omega_k(t) - iP_k(t)]a_k(t) = 0, \tag{9.11}$$

$$\bar{a}_k[\overleftarrow{\frac{d}{dt}} - i\omega_k(t) + iP_k(t)] = 0, \tag{9.12}$$

where

$$P_k(t) \equiv iB_k^{-1}(t)\frac{d}{dt}B_k(t). \tag{9.13}$$

It is due to this definition of $P_k(t)$ that $[d/dt - iP_k(t)]$ is called *covariant time derivative* [7]. This suggests that one could develop this approach along the lines of a gauge theory, but we do not consider this point here.

The equations in (9.12) show that $\bar{a}_k(t)a_k(t)$ is independent of time. Note that this is not the number operator which is given by $a_k^\dagger(t)a_k(t) = (\bar{a}_k(t)\tau_3 a_k(t))^{11}$.

The unperturbed Hamiltonian for the a-operators is

$$\hat{H}_Q(t) = \hat{H}_0(t) - \hat{Q}(t), \tag{9.14}$$

where

$$\hat{H}_0(t) = \int d^3k\, \bar{a}_k(t)\omega_k(t)a_k(t) \tag{9.15}$$

and

$$\hat{Q}(t) = \int d^3k\, \bar{a}_k(t)P_k(t)a_k(t), \tag{9.16}$$

because this gives

$$i\frac{d}{dt}a_k(t) = [a_k(t), \hat{H}_Q], \tag{9.17}$$

$$i\frac{d}{dt}\bar{a}_k(t) = [\bar{a}_k(t), \hat{H}_Q]. \tag{9.18}$$

The time-dependent Bogoliubov matrix $B_k(t)$ is obtained from $B_k(\theta)$ by replacing θ by time. Recall the expression of $B_k(\theta)$ in (7.86) which was expressed in terms of three parameters (α_k, s_k, n_k); the parameter θ stands for these three parameters. In general we can make these parameters depend on time. Then, a thermal time-behavior is represented by a trajectory in the three dimensional space of $n_k(t)$, $\alpha_k(t)$ and $s_k(t)$ for which we observe the SU(1,1)-algebra. An extension of this to SU(1,1) gauge theory has been studied by Henning and his collaborators. We leave this interesting development to the paper [8]. Here we use the simple choice for α and s_k which was used in the t-representation formalism developed in the last chapter: $\alpha = 1$ and $s_k(t) = \ln[1 + \sigma n_k(t)]^{1/2}$.

Then, according to (8.95) we have

$$B_k(t) = \left[\begin{array}{cc} 1 + \sigma n_k(t) & -n_k(t) \\ -\sigma & 1 \end{array} \right], \qquad (9.19)$$

the inverse of which is

$$B_k^{-1}(t) = \left[\begin{array}{cc} 1 & n_k(t) \\ \sigma & 1 + \sigma n_k(t) \end{array} \right]. \qquad (9.20)$$

The linear nature of $B_k(t)$ in $n_k(t)$ frequently simplifies calculations.

The matrix P_k now takes the simple form

$$P_k(t) = i\sigma \dot{n}_k(t) T_0 \qquad (9.21)$$

and the operator $\hat{Q}$ in (9.16) becomes

$$\hat{Q}(t) = i\sigma \int d^3k \, \dot{n}_k(t) \bar{a}_k(t) T_0 a_k(t). \qquad (9.22)$$

Here T_0 is

$$T_0 \equiv \left[\begin{array}{cc} 1 & -\sigma \\ \sigma & -1 \end{array} \right], \qquad (9.23)$$

as was defined in the chapter 8.

From these relations it follows that *the unperturbed Hamiltonian changes its sign under tilde conjugation:*

$$[\hat{H}_Q(t)]^\sim = -\hat{H}_Q(t). \qquad (9.24)$$

Let us now turn to a case with interactions. As mentioned, when a system is given, its Lagrangian is specified. From this the dynamical Hamiltonian H easily follows according to the Hamilton-Jacobi formalism. This is not the Hamiltonian in TFD but the dynamical energy operator. The H has a free part and an interaction part:

$$H = H_0 + H_{int}. \qquad (9.25)$$

Then the TFD Hamiltonian is

$$\hat{H} = H - \tilde{H} = \hat{H}_0 + \hat{H}_{int} \qquad (9.26)$$

with

$$\hat{H}_{int} = H_{int} - \tilde{H}_{int}. \tag{9.27}$$

In the interaction representation for perturbative calculations, the operator change is generated by the unperturbed Hamiltonian, while time change in states is generated by the interaction Hamiltonian $\hat{H}_I$. The unperturbed Hamiltonian is not $\hat{H}_0$ but the $\hat{H}_Q$ obtained in the last subsection:

$$\hat{H}_Q = \hat{H}_0 - \hat{Q}. \tag{9.28}$$

The $\hat{Q}$-term is the modification of the unperturbed Hamiltonian due to the time-dependent Bogoliubov matrix. Since the total Hamiltonian remains as $\hat{H}$, the interaction Hamiltonian $\hat{H}_I$ contains the $\hat{Q}$-counter term:

$$\hat{H}_I = \hat{H}_{int} + \hat{Q}. \tag{9.29}$$

So far the particle number has been defined through the number parameter associated with the Bogoliubov transformation. Since the quasi-particle picture should include the interaction effects, we need to modify this particle number because of the interactions. We do this by studying corresponding modification of the Bogoliubov transformation. This gives the corrected number of quasi-particles. On the other hand the observed particle number can be obtained from the vacuum expectation value of the number operator in the Heisenberg representation. This is called the Heisenberg number. It is important that the two numbers should agree within the energy uncertainty caused by the thermal instability. In later sections we will find that *the corrected number parameter in the Bogoliubov transformation does coincide with the Heisenberg number* when a number fluctuation effect is separated out.

9.2.2 THE UNPERTURBED ONE BODY PROPAGATOR

In this section we study the unperturbed propagator for the fields controlled by the unperturbed thermal Hamiltonian $\hat{H}_Q$.

The unperturbed propagator is

$$\Delta_k(t, t')^{\mu\nu} \delta(\vec{k} - \vec{l}) = -i\langle 0|T[a_k(t)^{\mu} \bar{a}_l(t)^{\nu}|0\rangle, \tag{9.30}$$

where the oscillator operators a_k and $\bar{a}_k$ are the ones introduced in (9.1) and (9.2) through the time-dependent Bogoliubov transformation. Therefore, we have the following unperturbed propagator [5, 7, 6]:

$$\Delta_k(t, t')^{\mu\nu} = \left[B_k^{-1}(t) D_k(t, t') B_k(t') \right]^{\mu\nu}, \tag{9.31}$$

where $D_k(t, t')$ is the diagonal matrix defined by

$$D_k(t, t') = \begin{bmatrix} -i\theta(t - t') & 0 \\ 0 & i\theta(t' - t) \end{bmatrix} e^{-i \int_{t'}^{t} ds\, \omega_k(s)}. \tag{9.32}$$

This has a remarkable structure in that the propagator is given by the diagonal propagator sandwiched between the Bogoliubov matrices. When this propagator is used for the lines in Feynman diagrams, the essential part of the thermal change is taken care of at each vertex and the propagation wave propagates with no thermal mixing. We are going to find in later sections that this structure of the one body propagator is preserved with a slight modification even when we consider interaction effects.

So far we have described the unperturbed propagator for time-dependent thermal phenomena in terms of the oscillator operator $a_k(t)$. We now translate this result in terms of the free field φ. For simplicity, we consider a field of type 1:

$$\varphi(x)^\mu = \int \frac{d^3k}{(2\pi)^{3/2}} e^{i\vec{k}\cdot\vec{x} - i\int_{-\infty}^t \omega_k(s)ds} a_k(t)^\mu$$

$$\bar{\varphi}(x)^\mu = \int \frac{d^3k}{(2\pi)^{3/2}} e^{-i\vec{k}\cdot\vec{x} + i\int_{-\infty}^t \omega_k(s)ds} \bar{a}_k(t)^\mu, \qquad (9.33)$$

because it is straightforward to extend this to fields of type 2.

The propagator is defined by

$$\Delta_c(x, x')^{\mu\nu} = -i\langle 0|T[\varphi(x)^\mu \bar{\varphi}(x')^\nu]|0\rangle. \qquad (9.34)$$

This can be computed by means of the above result as

$$\Delta_c(x - x')^{\mu\nu} = \int \frac{d^3k}{(2\pi)^3} e^{i\vec{k}\cdot(\vec{x}-\vec{x}') - i\int_{t'}^t \omega_k(s)ds} \Delta_k(t, t'). \qquad (9.35)$$

This is the unperturbed propagator, which is used for the internal Feynman line.

9.2.3 THE CORRECTED PARTICLE NUMBER

In this subsection we study the one body propagator corrected by interactions.

We consider a Heisenberg field of type 1:

$$\psi(x)^\mu = \int \frac{d^3k}{(2\pi)^{3/2}} e^{i\vec{k}\cdot\vec{x}} a_{Hk}(t)^\mu$$

$$\bar{\psi}(x)^\mu = \int \frac{d^3k}{(2\pi)^{3/2}} e^{-i\vec{k}\cdot\vec{x}} \bar{a}_{Hk}(t)^\mu. \qquad (9.36)$$

The propagator is defined by

$$G_c(x - x')^{\mu\nu} = -i\langle 0|T[\psi(x)^\mu \bar{\psi}(x')^\nu]|0\rangle. \qquad (9.37)$$

We put this in the three dimensional Fourier form:

$$G_c(x - x')^{\mu\nu} = -i\int \frac{d^3k}{(2\pi)^3} e^{i\vec{k}\cdot(\vec{x}-\vec{x}')} G_k(t, t')^{\mu\nu}, \qquad (9.38)$$

where $G_k(t, t')^{\mu\nu}$ is defined by

$$G_k(t, t')^{\mu\nu} \delta(\vec{k} - \vec{l}) = -i\langle 0|T[a_{Hk}(t)^\mu \bar{a}_{Hl}(t')^\nu]|0\rangle. \qquad (9.39)$$

Rewriting this quantity in terms of operators in the interaction representation, we have

$$G_k(t, t')^{\mu\nu} \delta(\vec{k} - \vec{l}) = -i\langle 0|T[a_k(t)^\mu \bar{a}_l(t')^\nu \hat{S}]|0\rangle. \qquad (9.40)$$

Here we use the interaction representation with $\hat{H}_Q(t)$ as the unperturbed Hamiltonian and $\hat{H}_I$ as the interaction Hamiltonian. The operator $\hat{S}$ is introduced as

$$i\frac{d}{dt}\hat{u}(t, t_0) = \hat{H}_I(t)\hat{u}(t, t_0), \qquad (9.41)$$

$$\hat{u}(t_0, t_0) = 1 \qquad (9.42)$$

and

$$\hat{S} \equiv \hat{u}(\infty, -\infty) \qquad (9.43)$$

$$= T\left[\exp[-i\int_{-\infty}^{\infty} ds\, \hat{H}_I(s)]\right]. \qquad (9.44)$$

As was mentioned previously we use $t_0 = -\infty$. Here t_0 is the time when the Heisenberg and interaction representations coincide.

Then, we have the following relation:

$$F_k(t, t')^{\mu\nu} \delta(\vec{k} - \vec{l})$$
$$\equiv B[n_k(t)]^{\mu\mu'} G_k(t, t')^{\mu'\nu'} B^{-1}[n_k(t')]^{\nu'\nu} \delta(\vec{k} - \vec{l})$$
$$= -i\theta(t - t')\langle 0|\xi_k(t)^\mu \hat{u}(t, t')\bar{\xi}_l(t')^\nu \hat{u}(t', -\infty)|0\rangle$$
$$+ i\theta(t' - t)\langle 0|\bar{\xi}_l(t')^\nu \hat{u}(t', t)\xi_k(t)^\mu \hat{u}(t, -\infty)|0\rangle. \qquad (9.45)$$

Here we used the relation

$$\langle 0|\hat{u}(\infty, t) = \langle 0|, \qquad (9.46)$$

which results from the choice $\alpha = 1$, giving

$$\langle 0|\hat{H}_I(t) = 0. \qquad (9.47)$$

Since ξ_k^2 and $\bar{\xi}_k^1$ annihilate the bra-vacuum, we have

$$F_k(t, t')^{21} = 0,$$
$$F_k(t, t')^{11} = -i\theta(t - t')g(t, t' : \vec{k})^{11},$$
$$F_k(t, t')^{22} = i\theta(t' - t)g(t, t' : \vec{k})^{22}, \qquad (9.48)$$

while F_k^{12} does not vanish in general,

$$F_k(t, t')^{12} \equiv g(t, t' : \vec{k})^{12} \neq 0. \tag{9.49}$$

Here $g(t, t' : \vec{k})^{11}$, $g(t, t' : \vec{k})^{22}$ and $g(t, t' : \vec{k})^{12}$ are functions which are expressed as the integrals of internal times $\{s_i\}$ with integration ranges from $-\infty$ to t, t' and $\max\{t, t'\}$, respectively.

In this way we obtain the general structure of $G_k(t, t')^{\mu\nu}$ in (9.40) as

$$\begin{aligned}
G_k(t, t')^{\mu\nu} = B^{-1}&[n_k(t)]^{\mu\mu'} \\
\times &\begin{bmatrix} -i\theta(t - t')g(t, t' : \vec{k})^{11} & g(t, t' : \vec{k})^{12} \\ 0 & i\theta(t' - t)g(t, t' : \vec{k})^{22} \end{bmatrix}^{\mu'\nu'} B[n_k(t')]^{\nu'\nu}.
\end{aligned} \tag{9.50}$$

Let us introduce $g_\pm(t, t' : \vec{k})$ as

$$g(t, t' : \vec{k})^{12} = -i\theta(t - t')g_+(t, t' : \vec{k}) + i\theta(t' - t)g_-(t, t' : \vec{k}). \tag{9.51}$$

When the tilde conjugation rules given in subsection 7.2.2 applied to (9.39), we find

$$g^*(t, t' : \vec{k})^{11} = g(t', t : \vec{k})^{22} \tag{9.52}$$

$$g_+^*(t, t' : \vec{k}) = -g_-(t', t : \vec{k}). \tag{9.53}$$

Now we show that the definition of the number density parameter in the Bogoliubov matrix is not unique at this point. To see this we introduce the following elementary matrix relation:

$$D_2 \begin{bmatrix} a & b \\ 0 & c \end{bmatrix} D_1^{-1} = \begin{bmatrix} a & b - ad_1 + cd_2 \\ 0 & c \end{bmatrix}, \tag{9.54}$$

which becomes diagonal when we choose d_1 and d_2 as

$$b - ad_1 + cd_2 = 0. \tag{9.55}$$

In the above the matrices D_i are

$$\begin{bmatrix} 1 & d_i \\ 0 & 1 \end{bmatrix}. \tag{9.56}$$

Furthermore, since we have the relation

$$B[n]B^{-1}[N] = \begin{bmatrix} 1 & N - n \\ 0 & 1 \end{bmatrix}, \tag{9.57}$$

the matrices D_i can be formed by the Bogoliubov matrix.

This shows that we can change the number density parameter in the Bogoliubov transformation by modifying $g(t, t' : \vec{k})^{12}$ in (9.50). Also the above argument shows that we can diagonalize the propagator appearing between the Bogoliubov matrices in (9.50) by choosing suitable number density parameters. In brief, the number density parameters in the Bogoliubov transformation are uniquely determined by the diagonal propagation requirement which states that the one body propagator should have a form in which a diagonal propagator is sandwiched between two Bogoliubov matrices.

Using this diagonalization method we can rewrite (9.50) as

$$
G_k(t, t')^{\mu\nu} = B^{-1}[N_{2,k}(t, t')]^{\mu\mu'}
$$
$$
\times \begin{bmatrix} -i\theta(t - t')g(t, t' : \vec{k})^{11} & 0 \\ 0 & i\theta(t' - t)g(t, t' : \vec{k})^{22} \end{bmatrix}^{\mu'\nu'} B[N_{1,k}(t, t')]^{\nu'\nu},
$$
(9.58)

where the number density parameters are

$$
N_{1,k}(t, t') = n_k(t') - i\frac{g(t, t' : \vec{k})^{12}}{g(t, t' : \vec{k})^{11}}
$$
$$
N_{2,k}(t, t') = n_k(t) - i\frac{g(t, t' : \vec{k})^{12}}{g(t, t' : \vec{k})^{22}}.
$$
(9.59)

The diagonal propagation condition requires the above form of $N_{1,k}$ $(N_{2,k})$ for $t \geq t'$ $(t' \geq t)$. However we have extended this range to the entire time domain.

We may rewrite $N_{i,k}$ thus determined as

$$
N_{1,k}(t, t') = N_{R1,k}(t') + \nu_{1,k}(t, t')
$$
$$
N_{2,k}(t, t') = N_{R2,k}(t) + \nu_{2,k}(t, t')
$$
(9.60)

in such a way that

$$
\nu_{j,k}(t, t) = 0 \qquad j = 1, 2.
$$
(9.61)

$N_{Rj,k}$ and $\nu_{j,k}$ are expressed in terms of g-functions as

$$
N_{R1,k}(t) = n_k(t) - \frac{g_+(t, t : \vec{k})}{g(t, t : \vec{k})^{11}}
$$
(9.62)

$$
N_{R2,k}(t) = n_k(t) + \frac{g_-(t, t : \vec{k})}{g(t, t : \vec{k})^{22}}
$$
(9.63)

$$
\nu_{1,k}(t, t') = -\left(\frac{g_+(t, t' : \vec{k})}{g(t, t' : \vec{k})^{11}} - \frac{g_+(t', t' : \vec{k})}{g(t', t' : \vec{k})^{11}} \right)
$$
(9.64)

$$
\nu_{2,k}(t, t') = \frac{g_-(t, t' : \vec{k})}{g(t, t' : \vec{k})^{22}} - \frac{g_-(t, t : \vec{k})}{g(t, t : \vec{k})^{22}}.
$$
(9.65)

It is easy to see the following relations from the properties of g-functions under the tilde conjugation, (9.52) and (9.53),

$$N_{1,k}^*(t,t') = N_{2,k}(t',t) \tag{9.66}$$

$$N_{R1,k}^*(t) = N_{R2,k}(t) \tag{9.67}$$

$$\nu_{1,k}^*(t,t') = \nu_{2,k}(t',t) \tag{9.68}$$

The number density parameter $N_{R1,k}(t)$ (and $N_{R2,k}(t)$) is called the corrected number. The significance of this number will be found in the next subsection.

9.2.4 HEISENBERG NUMBER AND CORRECTED NUMBER

We now show that the corrected number appearing in the Bogoliubov transformation is the observed particle number, which is given by the Heisenberg number. Here the Heisenberg number $n_{Hk}(t)$ is defined by the Heisenberg operators as

$$n_{Hk}(t)\delta(\vec{k} - \vec{l}) = \langle 0|\bar{a}_{Hk}(t)^1 a_{Hl}(t)^1|0\rangle. \tag{9.69}$$

To show this, we recall the definition of $G_k(t,t')^{\mu\nu}$ in terms of the Heisenberg operators and take the limit of $t \longrightarrow t'$. This limit with $t \geq t'$ gives

$$1 + n_{Hk}(t) = g(t,t:\vec{k})^{11}[1 + N_{R1,k}(t)], \tag{9.70}$$

while the same limit with $t' \geq t$ leads to

$$n_{Hk}(t) = g(t,t:\vec{k})^{22} N_{R2,k}(t). \tag{9.71}$$

When the relations $g^*(t,t:\vec{k})^{11} = g(t,t:\vec{k})^{22}$ and $N_{R1,k}^* = N_{R2,k}$ are considered, the above two relations give

$$g(t,t:\vec{k})^{11} = g(t,t:\vec{k})^{22} = 1, \tag{9.72}$$

which with the above two equations leads to

$$N_{R1,k}(t) = N_{R2,k}(t) = n_{Hk}(t). \tag{9.73}$$

This shows that *the corrected Bogoliubov number parameter naturally becomes the Heisenberg particle number* when the diagonal propagation condition is enforced. This justifies the diagonal propagation condition.

Here is a physical implication of this result. The full propagator $G_k(t,t')^{\mu\nu}$ has the matrix structure of $B^{-1} \times$ (a diagonal matrix) $\times B$, which is similar to that of the unperturbed propagator $\Delta_k(t,t')^{\mu\nu}$ in (9.31). However, the unperturbed B-matrices are parameterized by the unperturbed number $n(t)$. On the other hand the number density parameters in the B-matrices corrected by interactions are more complicated, given by $N_{i,k}(t,t')$ ($i = 1, 2$), which depend on two times. However, as (9.60) shows, they are separated into two parts. One, depending on a single time, turned out to be the Heisenberg number density $n_{Hk}(t)$. The other, denoted by $\nu_{j,k}(t,t')$, is interpreted as the fluctuation effect, because this vanishes for $t = t'$. Thus we have

$$N_{j,k}(t,t') = n_{Hk}(t) + \text{fluctuation effects.} \qquad (9.74)$$

Summarizing, we apply the diagonal propagation condition to every order of a perturbative calculation. We then obtain the Bogoliubov number parameter which depends on two times. Taking the two times to be equal, we obtain the corrected number, which is guaranteed to be same as the Heisenberg number [1, 2].

This approach eliminates an unpleasant aspect of TFD. In the past, we had two kinds of number. One is the Heisenberg number, while the other is the Bogoliubov number parameter. In this section we have reconciled these two kinds of number, by showing that the corrected number parameter is the Heisenberg number.

9.2.5   PRODUCT RULES

In the last subsection we saw that the one body propagators corrected by interactions have the form

$$B^{-1}[N^2(t,t')](\text{diagonal matrix})B[N^1(t,t')] \qquad (9.75)$$

with

$$N^2(t,t') = n_H(t') + \text{fluctuation,} \qquad (9.76)$$
$$N^1(t,t') = n_H(t) + \text{fluctuation.} \qquad (9.77)$$

Here the momentum suffix was omitted.

In this subsection we will show that self-energy diagrams consisting of lines connecting two points have the same structure with *no fluctuations*. This is the product rule in the t-representation for time-dependent thermal situations. Note that this simplicity is easily destroyed as soon as we include other internal vertices. Still these diagrams are useful in illustrating the calculation of self-energy diagrams, because they are relatively simple even though they include interaction effects.

Thus we consider the type of self-energy Feynman diagrams, shown in the Figure 3. Any self-energy diagram with only two vertices has this structure.

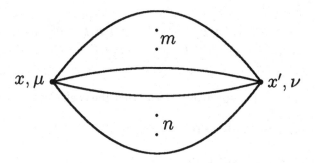

Figure 3. The self-energy diagram for which the product rule holds.

The general expression for these self-energy Feynman diagrams, $\Sigma_{m,n}(x,x')^{\mu\nu}$, is given by

$$
\begin{aligned}
\Sigma_{m,n}(x,x')^{11} &= i^{m+n-1}C(\Delta(x,x')^{11})^m(\Delta(x',x)^{11})^n, \\
\Sigma_{m,n}(x,x')^{12} &= i^{m+n-1}C(-1)^{m+1}(\Delta(x,x')^{12})^m(\Delta(x',x)^{21})^n, \\
\Sigma_{m,n}(x,x')^{21} &= i^{m+n-1}C(-1)^{n}(\Delta(x,x')^{21})^m(\Delta(x',x)^{12})^n, \\
\Sigma_{m,n}(x,x')^{22} &= i^{m+n-1}C(-1)^{m+n+1}(\Delta(x,x')^{22})^m(\Delta(x',x)^{22})^n,
\end{aligned}
$$

$$(9.78)$$

C being a positive number. Here m and n are positive integers. Remember that the propagator $\Delta(x,x')^{\mu\nu}$ is the unperturbed propagator whose spatial Fourier transform is the $\Delta_k(t,t')$ given in (9.31).

We study the one loop self-energy diagrams with $(m = 2, n = 0)$ or $(m = 1, n = 1)$. (See [2] for general consideration.) The self-energy term in (9.78) becomes in the Fourier space with respect to $\vec{x} - \vec{x'}$

$$
\begin{aligned}
&\Sigma_{2,0}(t,t',\vec{k})^{\mu\nu} \\
&= iC \int \frac{d^3q}{(2\pi)^3} \\
&\quad \times \begin{bmatrix} \Delta(t,t',\vec{k}-\vec{q})^{11}\Delta(t,t',\vec{q})^{11} & -\Delta(t,t',\vec{k}-\vec{q})^{12}\Delta(t,t',\vec{q})^{12} \\ \Delta(t,t',\vec{k}-\vec{q})^{21}\Delta(t,t',\vec{q})^{21} & -\Delta(t,t',\vec{k}-\vec{q})^{22}\Delta(t,t',\vec{q})^{22} \end{bmatrix}
\end{aligned}
$$

$$(9.79)$$

for $m = 2$ and $n = 0$, and

$$\Sigma_{1,1}(t, t', \vec{k})^{\mu\nu}$$

$$= iC \int \frac{d^3q}{(2\pi)^3}$$

$$\times \begin{bmatrix} \Delta(t, t', \vec{k} - \vec{q})^{11}\Delta(t', t, -\vec{q})^{11} & \Delta(t, t', \vec{k} - \vec{q})^{12}\Delta(t', t, -\vec{q})^{21} \\ -\Delta(t, t', \vec{k} - \vec{q})^{21}\Delta(t', t, -\vec{q})^{12} & -\Delta(t, t', \vec{k} - \vec{q})^{22}\Delta(t', t, -\vec{q})^{22} \end{bmatrix}$$

$$(9.80)$$

for $m = 1$ and $n = 1$, respectively. Recall that the coefficient C is a positive numerical factor.

Each matrix element is written as

$$\Delta(t, t', \vec{k} - \vec{q})^{\mu\nu}\Delta(t, t', \vec{q})^{\mu\nu}$$

$$= D_{k-q}(t, t')^{11}D_q(t, t')^{11}$$

$$\times \left[B_{k-q}^{-1}(t)\frac{1 + \tau_3}{2}B_{k-q}(t') \right]^{\mu\nu} \left[B_q^{-1}(t)\frac{1 + \tau_3}{2}B_q(t') \right]^{\mu\nu}$$

$$+ D_{k-q}(t, t')^{22}D_q(t, t')^{22}$$

$$\times \left[B_{k-q}^{-1}(t)\frac{1 - \tau_3}{2}B_{k-q}(t') \right]^{\mu\nu} \left[B_q^{-1}(t)\frac{1 - \tau_3}{2}B_q(t') \right]^{\mu\nu}$$

$$(9.81)$$

$$\Delta(t, t', \vec{k} - \vec{q})^{\mu\nu}\Delta(t', t, -\vec{q})^{\nu\mu}$$

$$= D_{k-q}(t, t')^{11}D_{-q}(t', t)^{22}$$

$$\times \left[B_{k-q}^{-1}(t)\frac{1 + \tau_3}{2}B_{k-q}(t') \right]^{\mu\nu} \left[B_{-q}^{-1}(t')\frac{1 - \tau_3}{2}B_{-q}(t) \right]^{\nu\mu}$$

$$+ D_{k-q}(t, t')^{22}D_{-q}(t', t)^{11}$$

$$\times \left[B_{k-q}^{-1}(t)\frac{1 - \tau_3}{2}B_{k-q}(t') \right]^{\mu\nu} \left[B_{-q}^{-1}(t')\frac{1 + \tau_3}{2}B_q(t) \right]^{\nu\nu}$$

$$(9.82)$$

according to (9.31) together with (9.32). Here the dummy indices μ and ν are not summed over.

According to (8.110) and (8.111) we have

$$B^{-1}(t)\frac{1 + \tau_3}{2}B(t') = T_+ + n(t')T_0$$

$$B^{-1}(t)\frac{1 - \tau_3}{2}B(t') = T_- - n(t)T_0. \qquad (9.83)$$

Thus we rewrite (9.81) and (9.82) as

$$\Delta(t, t', \vec{k} - \vec{q})^{\mu\nu} \Delta(t, t', \vec{q})^{\mu\nu}$$

$$= D_{k-q}(t, t')^{11} D_q(t, t')^{11} \frac{n_{k-q}(t') n_q(t')}{N_+(t')}$$

$$\times \quad [\{T_+ + N_+(t') T_0\} \tau_3]^{\mu\nu}$$

$$- D_{k-q}(t, t')^{22} D_q(t, t')^{22} \frac{n_{k-q}(t) n_q(t))}{N_+(t)}$$

$$\times \quad [\{T_- - N_+(t) T_0\} \tau_3]^{\mu\nu}, \tag{9.84}$$

$$\Delta(t, t', \vec{k} - \vec{q})^{\mu\nu} \Delta(t', t, -\vec{q})^{\nu\mu}$$

$$= -D_{k-q}(t, t')^{11} D_{-q}(t', t)^{22} \frac{n_{k-q}(t')(1 + n_{-q}(t'))}{N_-(t')}$$

$$\times \quad [\tau_3 \{T_+ + N_-(t') T_0\}]^{\mu\nu}$$

$$+ D_{k-q}(t, t')^{22} D_{-q}(t', t)^{11} \frac{n_{k-q}(t)(1 + n_{-q}(t))}{N_-(t)}$$

$$\times \quad [\tau_3 \{T_- - N_-(t) T_0\}]^{\mu\nu}, \tag{9.85}$$

where

$$N_\pm(t) \equiv \frac{F_\pm(t)}{1 - F_\pm(t)} \tag{9.86}$$

with

$$F_+(t) \quad \equiv F_+(t : \vec{k} - \vec{q}, \vec{q}) \quad = f_{k-q}(t) f_q(t) \tag{9.87}$$

$$F_-(t) \quad \equiv F_-(t : \vec{k} - \vec{q}, -\vec{q}) \quad = f_{k-q}(t) f_{-q}^{-1}(t), \tag{9.88}$$

while the function f_k is related to n_k through

$$n_k(t) = \frac{f_k(t)}{1 - f_k(t)}. \tag{9.89}$$

The above results for $N_\pm$ has a remarkable property. The relations (9.87) and (9.88) show that, when f_k has the form $\exp[-\beta(t)\omega_k]$, $F_\pm$ is $\exp[-\beta(t)(\omega_{k-q} \pm \omega_q)]$, implying that, when $n_k(t)$ has the Boltzmann form, so does $N_\pm$. This result holds true for any choice of (m, n).

Recall now the definition (9.32). Equations (9.84) and (9.85) have the form given on the right hand side of (8.113), so use of (9.83) leads to

$$\Sigma_{2,0}(t, t', \vec{k})^{\mu\nu} = C \int \frac{d^3q}{(2\pi)^3} \left[B^{-1}[N_+(t)] \right.$$

$$\times \quad \begin{bmatrix} -i\theta(t - t') s_{2,0}(t') & 0 \\ 0 & i\theta(t' - t) s_{2,0}(t) \end{bmatrix} B[N_+(t')] \Big]^{\mu\nu}$$

$$\times \quad e^{-i \int_{t'}^{t} ds(\omega_{k-q}(s) + \omega_q(s))}, \tag{9.90}$$

$$\Sigma_{1,1}(t,t',\vec{k})^{\mu\nu} = C \int \frac{d^3q}{(2\pi)^3} \left[B^{-1}[N_-(t)] \right.$$

$$\times \left. \begin{bmatrix} -i\theta(t-t')s_{1,1}(t') & 0 \\ 0 & i\theta(t'-t)s_{1,1}(t) \end{bmatrix} B[N_-(t')] \right]^{\mu\nu}$$

$$\times \; e^{-i\int_{t'}^{t} ds(\omega_{k-q}(s)-\omega_{-q}(s))}. \tag{9.91}$$

Here we used the relations

$$s_{2,0}(t) \;=\; \frac{n_{k-q}(t)n_q(t))}{N_+(t)}, \tag{9.92}$$

$$s_{1,1}(t) \;=\; \frac{n_{k-q}(t)(1+n_{-q}(t))}{N_-(t)}, \tag{9.93}$$

$$N_+(t) \;=\; \frac{n_{k-q}(t)n_q(t)}{1+n_{k-q}(t)+n_q(t)}, \tag{9.94}$$

$$N_-(t) \;=\; \frac{n_{k-q}(t)[1+n_{-q}(t)]}{n_{-q}(t)-n_{k-q}(t)}. \tag{9.95}$$

The number parameters $N_\pm(t)$ are the particle numbers corrected by respective self-energy diagrams.

Exercise 4 Derive the following expression,

$$\Sigma_{2,1}(t,t',\vec{k})^{\mu\nu} = \int \prod_{i=1}^{3} \left(\frac{d^3q_i}{(2\pi)^3} \right) (2\pi)^3 \delta(\vec{k}-\vec{q}_1-\vec{q}_2+\vec{q}_3)$$

$$\times \left[B^{-1}[N_{2,1}(t)] \begin{bmatrix} -i\theta(t-t')s(t') & 0 \\ 0 & i\theta(t'-t)s(t) \end{bmatrix} B[N_{2,1}(t')] \right]^{\mu\nu}$$

$$\times \; e^{-i\int_{t'}^{t} ds(\omega_{q_1}(s)+\omega_{q_2}(s)-\omega_{q_3}(s))}, \tag{9.96}$$

where

$$s(t) = C \frac{n_{q_1}(t)n_{q_2}(t)(1+n_{q_3}(t))}{N_{2,1}(t)} \tag{9.97}$$

and

$$N_{2,1} = \frac{n_{q_1}n_{q_2}(1+n_{q_3})}{(1+n_{q_1})(1+n_{q_2})n_{q_3}-n_{q_1}n_{q_2}(1+n_{q_3})}. \tag{9.98}$$

Hint: Calculate the matrix element $\Sigma_{1,1}(t,t',\vec{q}_2-\vec{q}_3)^{\mu\nu}\Delta(t,t',\vec{q}_3)^{\mu\nu}$, using the result (9.91).

9.2.6   THE KINETIC EQUATION

We have developed a calculable formulation of the time-dependent nonequilibrium TFD in the interaction representation. Now any expectation value or propagators can be expressed in terms of time-dependent parameters, i.e., quasi-particle energy $\omega_k(t)$ and number density parameter $n_k(t)$ which characterize the unperturbed representation, and the other parameters of models such as coupling constants. So far $\omega_k(t)$ and $n_k(t)$ remain unknown. Thus to determine them is vital for making theoretical predictions in the present formalism. However, use of time-independent unperturbed ω_k considerably simplifies perturbative calculations. Thus, although the general formulation in this and the next sections permits time-dependent unperturbed ω_k, in this subsection, we use time-independent unperturbed ω_k, for example $\omega_k = \omega_k(t_0)$ with $t_0 = -\infty$. The time-dependent $\omega_k(t)$ will then be obtained from the self-energy, as will be shown later.

However time-dependence of unperturbed n_k is of vital significance in order to include the effect of the thermal generator $\hat{Q}$ which is proportional to $\dot{n}_k$. Since all the quantities are functionals of $n_k(t)$, we need an equation which determines $n_k(t)$.

In this section we try to derive the equation controlling the temporal behavior of $n_k(t)$ by means of the self-energy renormalization condition.

To make our point clear let us consider the model with $\psi^\dagger\psi^\dagger\psi\psi$-type interaction. The second order self-energy $\Sigma^{\mu\nu}$ for the model is the two loop self-energy diagram (9.96) corresponding to Fig. 4 with $m = 2$ and $n = 1$.

We emphasize again that the unperturbed energy parameter ω_k is independent of time. This considerably simplifies (9.96) and gives the following expression of self-energy in a compact notation; suppressing the suffix $\{2, 1\}$ and denoting $N_{2,1}$ simply by N,

$$\Sigma(t, t', \vec{k})^{\mu\nu} = \int [dq] \left[B^{-1}[N(t)] \begin{bmatrix} 1 & 0 \\ 0 & s(t) \end{bmatrix} \right.$$
$$\left. \times V(t - t') \begin{bmatrix} s(t') & 0 \\ 0 & 1 \end{bmatrix} B[N(t')] \right]^{\mu\nu} \qquad (9.99)$$

with

$$[dq] = \prod_{i=1}^{3} \left(\frac{d^3 q_i}{(2\pi)^3} \right) (2\pi)^3 \delta(\vec{k} - \vec{q}_1 - \vec{q}_2 + \vec{q}_3), \qquad (9.100)$$

$$s(t) = C \frac{n_{q_1}(t) n_{q_2}(t) [1 + n_{q_3}(t)]}{N(t)}, \qquad (9.101)$$

$$V(t - t')^{\mu\nu} = \begin{bmatrix} -i\theta(t - t') & 0 \\ 0 & i\theta(t' - t) \end{bmatrix}^{\mu\nu} \exp[-iW(t - t')], \qquad (9.102)$$

$$W = \omega_{q_1} + \omega_{q_2} - \omega_{q_3}. \qquad (9.103)$$

Here $n_q(t)$ is the unperturbed number parameter. The coefficient C is a positive numerical factor for the model.

The diagonal matrix $V(t - t')^{\mu\nu}$ describes the wave propagation of the quasi-particle and has the Fourier form

$$V(t - t')^{\mu\nu} \equiv \int \frac{dk_0}{2\pi} e^{-ik_0(t-t')} V(k_0)^{\mu\nu} \qquad (9.104)$$

$$V(k_0)^{\mu\nu} = \left[\frac{1}{k_0 - W + i\epsilon\tau_3} \right]^{\mu\nu}. \qquad (9.105)$$

The $V(t - t')^{\mu\nu}$ is of the same structure as the self-energy itself in the usual field theory. We therefore apply the usual procedure to $V(t - t')^{\mu\nu}$ to extract the on-shell part from the whole self-energy in (9.99): Put k_0 on the shell, $k_0 = \omega_k$, in the $V(k_0)^{\mu\nu}$ in (9.104) to get

$$V^{(0)}(t - t')^{\mu\nu} = V(k_0 = \omega_k)^{\mu\nu} \delta(t - t'). \qquad (9.106)$$

Substituting this for $V(t - t')^{\mu\nu}$ in (9.99), we have

$$\Sigma(t, t', \vec{k})^{(0)\mu\nu} = \delta(t - t') \int [dq]\, s(t) \left[B^{-1}[N(t)] V(k_0 = \omega_k) B[N(t)] \right]^{\mu\nu}. \qquad (9.107)$$

This is the on-shell self-energy in the t-representation. We remark that t and t' in N and s are equal due to the presence of $\delta(t - t')$ coming from $V^{(0)}(t - t')^{\mu\nu}$.

The real part of (9.107) is

$$\Re\Sigma(t, t', \vec{k})^{(0)\mu\nu} = \delta(t - t') \int [dq]\, s(t) \left[B^{-1}[N(t)] \frac{P}{\omega_k - W} B[N(t)] \right]^{\mu\nu}. \qquad (9.108)$$

This part contributes to the time-dependent energy shift $\delta\omega_k(t)$ in addition to the energy shift due to the one loop self-energy, i.e., the tag diagram.

The imaginary part of (9.107) is

$$\Im\Sigma(t, t', \vec{k})^{(0)\mu\nu} = -i\delta(t - t') \int [dq]\, s(t) \pi\delta(\omega_k - W) A[N(t)]^{\mu\nu}, \qquad (9.109)$$

where we used (9.105) and the formula $\Im[1/(x + i\epsilon\tau_3)] = -i\tau_3\delta(x)$. The matrix A is $B^{-1}\tau_3 B$.

In addition to the above self-energy the $\hat{Q}$-term in the interaction Hamiltonian $\hat{H}_I = \hat{H}_{int} + \hat{Q}$ gives rise to a self-energy counter term. The total on-shell self-energy will be denoted by Σ^0_{tot}. The real part is not modified by the $\hat{Q}$-term, but the imaginary part of the total on-shell self-energy becomes

$$\Im\Sigma_{tot}(t, t', \vec{k})^{(0)\mu\nu} = \delta\Sigma(t, \vec{k})^{\mu\nu} \delta(t - t') \qquad (9.110)$$

with

$$\delta\Sigma(t, \vec{k})^{\mu\nu} = -i \left[-\dot{n}_k(t) T_0 + \int [dq]\, s(t) \pi\delta(\omega_k - W) A[N(t)] \right]^{\mu\nu}. \qquad (9.111)$$

This introduces the self-energy contribution $\delta\hat{H} = \bar{a}_k(t)\delta\Sigma(t,\vec{k})a_k(t)$ to the quasi-particle Hamiltonian. We require that, at any t, *this self-energy term in the quasi-particle Hamiltonian be of a diagonal form in terms of* ξ_k^μ *and* $\bar{\xi}_k^\mu$. This is the self-consistent renormalization condition. Note that this requirement determines the vacuum $|0\rangle$. Since $\delta\hat{H}$ should change its sign under tilde conjugation, its imaginary term cannot have a $\bar{\xi}\xi$-term; it can have $\bar{\xi}\tau_3\xi$ only. Due to the relation $\bar{\xi}_k\tau_3\xi_k = \bar{a}_k(t)A_k(t)a_k(t)$, this condition reads

$$i\left[-\dot{n}_k(t)T_0 + \int[dq]\, s(t)\pi\delta(\omega_k - W)A[N(t)]\right]^{\mu\nu} = i\eta_k(t)A_k(n(t))^{\mu\nu},$$

$$(9.112)$$

where $\eta_k(t)$ is a real number. This is the self-consistent renormalization condition. As was mentioned previously this condition is the same as the requirement that the quasi-particle Hamiltonian $\hat{H}_0 + \delta\hat{H}$ is diagonal in ξ.

To solve (9.112) we recall the relation (8.112), that is

$$A(n) = T_+ - T_- + 2nT_0. \qquad (9.113)$$

We then find that the relation (9.112) is satisfied by the single equation

$$\dot{n}_k(t) = -2\kappa_k(t)n_k(t) + 2\pi C \int[dq]\, \delta(\omega_k - W)n_{q_1}(t)n_{q_2}(t)[1 + n_{q_3}(t)],$$

$$(9.114)$$

with

$$\begin{aligned}
\eta_k(t) &= \kappa_k(t) \equiv \int[dq]\, s(t)\pi\delta(\omega_k - W) \\
&= \pi C \int[dq]\, \delta(\omega_k - W)\left\{[1 + n_{q_1}(t)][1 + n_{q_2}(t)]n_{q_3}(t) \right. \\
&\quad \left. - n_{q_1}(t)n_{q_2}(t)[1 + n_{q_3}(t)]\right\}.
\end{aligned} \qquad (9.115)$$

The $\kappa_k(t)$ is the time-dependent dissipative coefficient. Equation (9.114) is the kinetic equation for the Bogoliubov parameter $n_k(t)$ in the two loop self-energy approximation without vertex correction. As within this approximation *the kinetic equation has the structure of the Boltzmann equation*, we will call it Boltzmann equation. This clearly shows how the unperturbed number $n_k(t)$ is influenced by interactions.

Let us investigate the characteristics of the Boltzmann equation (9.114) a bit further. Multiplying by an arbitrary function ϕ_k of $\vec{k}$ on both side of (9.114), and integrating it with respect to $\vec{k}$, we have

$$\begin{aligned}
\int d^3k\, \phi_k \dot{n}_k(t) &= \frac{(2\pi)^7 C}{2} \int \prod_{i=0}^{3}\left(\frac{d^3q_i}{(2\pi)^3}\right)\delta(0 - 1 - 2 + 3) \\
&\quad \times (\phi_3 + \phi_2 - \phi_1 - \phi_0)\, n_0 n_1(1 + n_2)(1 + n_3),
\end{aligned} \qquad (9.116)$$

where we introduced the abbreviations $w_i = w_{q_i}$, $\phi_i = \phi_{q_i}$, $n_i = n_{q_i}(t)$, and

$$\delta(0 - 1 - 2 + 3) = \delta(\vec{q}_0 - \vec{q}_1 - \vec{q}_2 + \vec{q}_3)\,\delta(w_0 - w_1 - w_2 + w_3). \quad (9.117)$$

To obtain the expression of the right hand side of (9.116), we changed the integration variables. We notice immediately that, for $\phi_k = 1$, the right hand side of (9.116) disappears. Furthermore, using the property of delta function, $x\delta(x) = 0$, we see that, for $\phi_k = \vec{k}$ or w_k, the right hand side of (9.116) is equal to zero. These properties indicate that the number density $\mathcal{N}$, the momentum density $\vec{\mathcal{P}}$ and the energy density $\mathcal{E}$ are conserved quantities. Here, they are defined, respectively, by

$$\mathcal{N} \;=\; \int d^3k\, n_k(t), \qquad\qquad (9.118)$$

$$\vec{\mathcal{P}} \;=\; \int d^3k\, \vec{k}\, n_k(t), \qquad\qquad (9.119)$$

$$\mathcal{E} \;=\; \int d^3k\, w_k\, n_k(t). \qquad\qquad (9.120)$$

In order to prove the conservation of energy density (9.120), for example, just replace ϕ in (9.116) by w. The conservation of these quantities reflects the fact that the system is an isolated system.

Exercise 5 Derive the expression of the right hand side of (9.116).

Let us further check the behavior of the entropy density $S(t)$ defined by

$$S(t) = \int d^3k\,\{[1 + n_k(t)]\ln[1 + n_k(t)] - n_k(t)\ln n_k(t)\}. \quad (9.121)$$

Changing the integration parameters, we have

$$\begin{aligned}
\dot{S}(t) \;=\;& \frac{(2\pi)^7 C}{4} \int \prod_{i=0}^{3}\left(\frac{d^3q_i}{(2\pi)^3}\right)\delta(0 - 1 - 2 + 3)\\
&\times [n_0 n_1(1 + n_2)(1 + n_3) - (1 + n_0)(1 + n_1)n_2 n_3]\\
&\times \ln\frac{n_0 n_1(1 + n_2)(1 + n_3)}{(1 + n_0)(1 + n_1)n_2 n_3}.
\end{aligned} \qquad (9.122)$$

As $(x - y)\ln(x/y) \geq 0$ for positive x and y, the equation (9.122) shows that the entropy density always increases in time. It is remarkable that we do find a time-independent Fock space for quasi-particles in such a manner that the quasi particle Hamiltonian is diagonal at every time.

Exercise 6 Derive the expression of the right hand side of (9.122).

As an important aside: If we had chosen $t_0 = \infty$, which together with the Feynman diagram method requires $\alpha = 0$, then the sign of the right hand side of the equation (9.114) would have changed, leading to the incorrect result that the entropy would decrease in time. The correct choice, $\alpha = 1$ and $t_0 = -\infty$, corresponds to the choice of the path in the closed path formalism. A derivation of the Boltzmann equation in the closed path formalism has been given in [9]. Since the closed path formalism does not have the parameter s_k which played a vital role in the above consideration, it is not obvious how the derivation of the Boltzmann equation in TFD is related to the one in the closed path formalism.

Derivation of the Boltzmann equation from the statistical mechanics applied to quantum fields has attracted attention of many physicists and there is a vast number of publications on this subject. In this chapter we concentrated on a TFD treatment of this subject.

9.3   Time-Space Dependent TFD

9.3.1   THE BOGOLIUBOV TRANSFORMATIONS

At first glance, we might suppose that spatial dependence in thermal situation could be introduced by making the Bogoliubov matrix B_k in $a_k = B_k^{-1}\xi_k$ dependent on space. This would lead to $\vec{x}$-dependent a_k. However, this approach does not work, because the commutation relation

$$[a_k(t)^\mu, \bar{a}_l(t)^\nu] = \delta_{\mu\nu}\delta(\vec{k} - \vec{l}) \tag{9.123}$$

forbids a_k from depending on $\vec{x}$.

However, the formula

$$B^{-1}(n)B(N) = 1 - \sigma(n - N)T_0 \tag{9.124}$$

shows us how to introduce space dependent Bogoliubov transformations. Here $B(n)$ means the Bogoliubov matrix with the number parameter n. This and the following formulae hold true for both boson and fermion fields; $\sigma = 1(-1)$ for boson (fermion).

All the formulae in this section hold true both for $\alpha = 1$ or $\alpha = 0$. However, as was shown in the last section, we should use $t_0 = -\infty$, which enforces the choice $\alpha = 1$.

Now note the identity

$$\int d^3q \int d^3x \int d^3y \, e^{i[-(\vec{k}-\vec{q})\cdot\vec{x}+(\vec{l}-\vec{q})\cdot\vec{y}]}$$
$$\times \left(n[t, \vec{x} : \frac{\vec{k}+\vec{q}}{2}] - n[t, \vec{y} : \frac{\vec{l}+\vec{q}}{2}] \right) = 0. \tag{9.125}$$

This together with (9.124) leads to the formula

$$\frac{1}{(2\pi)^6} \int d^3q \int d^3x \int d^3y \, e^{i[-(\vec{k}-\vec{q})\cdot\vec{x}+(\vec{l}-\vec{q})\cdot\vec{y}]}$$

$$\times \left(B^{-1}(n[t,\vec{x}:\frac{\vec{k}+\vec{q}}{2}]) B(n[t,\vec{y}:\frac{\vec{l}+\vec{q}}{2}]) \right)^{\mu\nu} = \delta_{\mu\nu}\delta(\vec{k}-\vec{l}).$$

$$(9.126)$$

This shows that the Bogoliubov transformations

$$a_k(t)^\mu = \frac{1}{(2\pi)^3} \int d^3q \int d^3x \, e^{-i(\vec{k}-\vec{q})\cdot\vec{x}}$$

$$\times \quad B^{-1}(n[t,\vec{x}:\frac{\vec{k}+\vec{q}}{2}])^{\mu\nu}\xi_q(t)^\nu, \qquad (9.127)$$

$$\bar{a}_k(t)^\mu = \frac{1}{(2\pi)^3} \int d^3q \int d^3x \, e^{i(\vec{k}-\vec{q})\cdot\vec{x}}\bar{\xi}_q(t)^\nu$$

$$\times \quad B(n[t,\vec{x}:\frac{\vec{k}+\vec{q}}{2}])^{\nu\mu}, \qquad (9.128)$$

satisfy the commutation relation (9.123). In the above

$$\xi_q(t) \equiv \xi_q e^{-i\int^t ds\,\omega_q(s)}, \qquad (9.129)$$

$$\bar{\xi}_q(t) \equiv \bar{\xi}_q e^{i\int^t ds\,\omega_q(s)}. \qquad (9.130)$$

Since the vacuum operators, ξ_q and $\bar{\xi}_q$ are independent of time and space, so is their vacuum $|0\rangle$. Note that the quantum energy $\omega_k(t)$ does not depend on $\vec{x}$. This is because the quasi-particle unperturbed energy is given by the quantum energy in the asymptotic domain which is infinitely far away from the domain with spatial thermal variation. This situation is same as the one which we saw in case of states with macroscopic objects such as solitons (see chapter 6). However, interaction creates a self-energy which can depend on $\vec{x}$ and thus a $\vec{x}$-dependent quantum energy. The latter is not the unperturbed quantum energy, for being the asymptotic energy, the unperturbed energy includes only that part of the self-energy which is independent of $\vec{x}$. This is the energy renormalization.

It is remarkable that the above Bogoliubov transformations have an inverse. With the use of (9.124) we can prove that

$$\xi_k(t)^\mu = \frac{1}{(2\pi)^3} \int d^3q \int d^3x \, e^{-i(\vec{k}-\vec{q})\cdot\vec{x}}$$

$$\times \quad B(n[t,\vec{x}:\frac{\vec{k}+\vec{q}}{2}])^{\mu\nu}a_q(t)^\nu, \qquad (9.131)$$

$$\bar{\xi}_k(t)^\mu = \frac{1}{(2\pi)^3} \int d^3q \int d^3x \, e^{i(\vec{k}-\vec{q})\cdot\vec{x}}\bar{a}_q(t)^\nu$$

$$\times \quad B^{-1}(n[t,\vec{x}:\frac{\vec{k}+\vec{q}}{2}])^{\nu\mu}. \qquad (9.132)$$

We close this subsection with a formula for the number parameter. A calculation shows that

$$n[t, \vec{x} : \vec{k}] = \int d^3q\, e^{-i\vec{q}\cdot\vec{x}} \langle 0|\bar{a}_{k+\frac{q}{2}}(t)^1 a_{k-\frac{q}{2}}(t)^1|0\rangle. \tag{9.133}$$

9.3.2   THE UNPERTURBED HAMILTONIAN

The above Bogoliubov transformations lead to the field equations

$$\begin{aligned}
i\dot{a}_k(t)^\mu &= \omega_k(t)a_k(t)^\mu \\
&+ i\int d^3q\, P(t,\vec{k},\vec{q})T_0^{\mu\nu}a_q(t)^\nu, \tag{9.134}
\end{aligned}$$

$$\begin{aligned}
i\dot{\bar{a}}_k(t)^\mu &= -\omega_k(t)\bar{a}_k(t)^\mu \\
&- i\int d^3q\, \bar{a}_q(t)^\nu T_0^{\nu\mu} P(t,\vec{q},\vec{k}), \tag{9.135}
\end{aligned}$$

where

$$P(t,\vec{k},\vec{q}) = P_0(t,\vec{k},\vec{q}) + P_1(t,\vec{k},\vec{q}) \tag{9.136}$$

with

$$\begin{aligned}
P_0(t,\vec{k},\vec{q}) &= \frac{1}{(2\pi)^3}\int d^3x\, e^{-i(\vec{k}-\vec{q})\cdot\vec{x}} \\
&\times \dot{n}[t,\vec{x} : \frac{\vec{k}+\vec{q}}{2}], \tag{9.137}
\end{aligned}$$

$$\begin{aligned}
P_1(t,\vec{k},\vec{q}) &= i\frac{1}{(2\pi)^3}\int d^3x\, e^{-i(\vec{k}-\vec{q})\cdot\vec{x}} \\
&\times \{\omega_k(t) - \omega_q(t)\}n[t,\vec{x} : \frac{\vec{k}+\vec{q}}{2}]. \tag{9.138}
\end{aligned}$$

These equations can be put in the form

$$i\dot{a}_k(t)^\mu = [a_k(t)^\mu, \hat{H}_Q(t)], \tag{9.139}$$

$$i\dot{\bar{a}}_k(t)^\mu = [\bar{a}_k(t)^\mu, \hat{H}_Q(t)], \tag{9.140}$$

where the unperturbed Hamiltonian $\hat{H}_Q$ is given by

$$\hat{H}_Q(t) = \hat{H}_0(t) - \hat{Q}(t) \tag{9.141}$$

with

$$\hat{H}_0(t) = \int d^3k\, \omega_k(t)\bar{a}_k(t)^\mu a_k(t)^\mu \tag{9.142}$$

and

$$\hat{Q}(t) = -i\int d^3k\int d^3l\, P(t,\vec{k},\vec{l})\bar{a}_k(t)T_0 a_l(t) \tag{9.143}$$

It may be worth noting that

$$\hat{H}_0(t) + i\int d^3k \int d^3l P_1(t, \vec{k}, \vec{l})\bar{a}_k(t)T_0 a_l(t) = \int d^3k\, \omega_k(t)\bar{\xi}_k\xi_k, \quad (9.144)$$

which is diagonal in terms of vacuum operators.

9.3.3 TIME-SPACE DEPENDENT TFD FOR FIELDS

We now reformulate the time-space dependent TFD in the last subsections in terms of the free fields

$$\varphi(t, \vec{x})^\mu = \frac{1}{(2\pi)^{3/2}} \int d^3k\, a_k(t)^\mu e^{i\vec{k}\cdot\vec{x}} e^{-i\int ds\,\omega_k(s)}, \quad (9.145)$$

$$\bar{\varphi}(t, \vec{x})^\mu = \frac{1}{(2\pi)^{3/2}} \int d^3k\, \bar{a}_k(t)^\mu e^{-i\vec{k}\cdot\vec{x}} e^{i\int ds\,\omega_k(s)}. \quad (9.146)$$

We introduce also the quasi-particle fields:

$$\xi(t, \vec{x})^\mu = \frac{1}{(2\pi)^{3/2}} \int d^3k\, \xi_k^\mu e^{i\vec{k}\cdot\vec{x}} e^{-i\int ds\,\omega_k(s)}, \quad (9.147)$$

$$\bar{\xi}(t, \vec{x})^\mu = \frac{1}{(2\pi)^{3/2}} \int d^3k\, \bar{\xi}_k^\mu e^{-i\vec{k}\cdot\vec{x}} e^{i\int ds\,\omega_k(s)}. \quad (9.148)$$

Then, the thermal Bogoliubov transformations become

$$\varphi(t, \vec{x})^\mu = \frac{1}{(2\pi)^3} \int d^3y \int d^3q\, e^{i\vec{q}\cdot(\vec{x}-\vec{y})}$$
$$\times\ B^{-1}(n[t, \frac{\vec{x}+\vec{y}}{2} : \vec{q}])^{\mu\nu}\xi(t, \vec{y})^\nu, \quad (9.149)$$

$$\bar{\varphi}(t, \vec{x})^\mu = \frac{1}{(2\pi)^3} \int d^3y \int d^3q\, e^{-i\vec{q}\cdot(\vec{x}-\vec{y})}$$
$$\times\ \bar{\xi}(t, \vec{y})^\nu B(n[t, \frac{\vec{x}+\vec{y}}{2} : \vec{q}])^{\nu\mu}. \quad (9.150)$$

The field equations are

$$i\dot{\varphi}(t, \vec{x})^\mu = \omega(-i\nabla)\varphi(t, \vec{x})^\mu$$
$$+\ i\int d^3y\, P(t, \vec{x}, \vec{y})T_0^{\mu\nu}\varphi(t, \vec{y})^\nu, \quad (9.151)$$

$$i\dot{\bar{\varphi}}(t, \vec{x})^\mu = -\omega(i\nabla)\bar{\varphi}(t, \vec{x})^\mu$$
$$-\ i\int d^3y\, \bar{\varphi}(t, \vec{y})^\nu T_0^{\nu\mu} P(t, \vec{y}, \vec{x}), \quad (9.152)$$

where

$$P(t, \vec{x}, \vec{y}) = P_0(t, \vec{x}, \vec{y}) + P_1(t, \vec{x}, \vec{y}) \quad (9.153)$$

with

$$P_0(t, \vec{x}, \vec{y}) = \frac{1}{(2\pi)^3} \int d^3q \, e^{i(\vec{x}-\vec{y})\cdot\vec{q}}$$

$$\times \; \dot{n}[t, \frac{\vec{x}+\vec{y}}{2} : \vec{q}], \tag{9.154}$$

$$P_1(t, \vec{x}, \vec{y}) = i\frac{1}{(2\pi)^3} \int d^3q \, e^{i(\vec{x}-\vec{y})\cdot\vec{q}}$$

$$\times \{\omega(-i\nabla_x) - \omega(-i\nabla_y)\} n[t, \frac{\vec{x}+\vec{y}}{2} : \vec{q}]. \tag{9.155}$$

The vacuum expectation value of current is found to be

$$\langle 0| - \frac{i}{2}\bar{\varphi}(t, \vec{x}) \overleftrightarrow{\nabla} \varphi(t, \vec{x})|0\rangle = \int \frac{d^3k}{(2\pi)^3} \vec{k} \, n[t, \vec{x}, \vec{k}], \tag{9.156}$$

where

$$\overleftrightarrow{\nabla} = \overrightarrow{\nabla} - \overleftarrow{\nabla}. \tag{9.157}$$

This provides us with the intuitive interpretation that $n[t, \vec{x}, \vec{k}]$ is the Fourier component of the number density at space point $\vec{x}$ at time t.

9.3.4 THE UNPERTURBED ONE BODY PROPAGATOR

The unperturbed one body propagator follows from the thermal Bogoliubov transformation as follows:

$$\Delta_c(t, \vec{x} : t', \vec{x}') \equiv -i\langle 0|T[\varphi(t, \vec{x})\bar{\varphi}(t', \vec{x}')]|0\rangle \tag{9.158}$$

$$= -\frac{i}{(2\pi)^9} \int d^3k \int d^3k' \int d^3y \int d^3y' \int d^3l$$

$$\times \; e^{i\vec{k}\cdot(\vec{x}-\vec{y})} e^{-i\vec{k}'\cdot(\vec{x}'-\vec{y}')} e^{i\vec{l}\cdot(\vec{y}-\vec{y}')} e^{-i\int_{t'}^{t} ds \, \omega_l(s)}$$

$$\times \; B^{-1}(n[t, \vec{y}, \frac{\vec{k}+\vec{l}}{2}]) \begin{bmatrix} \theta(t-t') & 0 \\ 0 & -\theta(t'-t) \end{bmatrix}$$

$$\times \; B(n[t', \vec{x}', \frac{\vec{k}'+\vec{l}}{2}]), \tag{9.159}$$

which consists of a multiple integration of terms involving a diagonal matrix sandwiched between Bogoliubov matrices.

This represents the internal lines of Feynman diagrams of the perturbative calculations. Study of the corrected one body propagator and derivation of the Boltzmann equation for time-space behavior of the unperturbed number $n[t, \vec{x} : \vec{k}]$ from the self-consistent renormalization condition follows the steps taken in the study of time-dependent TFD in section 2. It has been confirmed that calculation of the self-energy diagram in second order perturbative approximation and extraction of the on-shell self-energy part creates the self-energy contribution to the quasi-particle Hamiltonian and that the self-consistent renormalization condition leads to the Boltzmann equation for inhomogeneous thermal processes.

9.4 Observed Particles and Quasi-Particles

The t-representation TFD disclosed a remarkable feature of the concept of quasi-particles. *The quasi-particle operators* $(\xi_k^\mu, \bar{\xi}_k^\mu)$ *are defined to be those, with which the Fock space is built.* In other words, the quasi-particle operators are those in terms of which the dynamical maps are expressed. The quasi-particles thus defined are not the same as the oscillator operators $(a_k(t), \bar{a}_k(t))$ in the interaction representation. The latter operators describe the observed particles.

The t-representation TFD indicated that the quasi-particles are stable. This seems to be a fundamental requirement for the quasi-particle operators. Any time-dependent deviation in the thermal behavior of observed particles from the quasi-particle behavior is taken care of by a time-dependent expression (dynamical map) of the oscillator operators in the interaction representation in terms of the quasi-particle operators. This expression is the time-dependent Bogoliubov transformation. This distinction between the observed particles and the quasi-particles requires some modification in the terminology used in equilibrium TFD. There we have frequently used the word "dissipative quasi-particle", which should be corrected as "dissipative observed particles". As was shown in equation (8.59) in subsection 8.1.3, the thermal dissipation of observed particles in equilibrium TFD can be treated by a time-dependent Bogoliubov matrix.

The above consideration answers the questions raised in subsection 8.1.3. The quasi-particle Hamiltonian is $\hat{H}_0$, while the unperturbed Hamiltonian in equilibrium TFD is $\hat{H}_{prop}$. However, when the dissipative effect is perturbative, we may use $\hat{H}_0$ instead. The unperturbed Hamiltonian in the time-dependent TFD is $(\hat{H}_{prop} - \hat{Q})$. Here too, we may choose $(\hat{H}_0 - \hat{Q})$ for the unperturbed Hamiltonian, when the dissipative effect is perturbative, as we did in this chapter.

9.5 Comments

Equilibrium TFD in the t-representation developed in chapter 8 showed that a single quasi-particle goes through states with thermal fluctuations. We may apply a similar argument to the center of mass coordinate of the two body particle propagator. The result is expected to show that a two particle state may also go through states of thermal fluctuations. This is consistent with the picture that there may appear strong thermal effects during high energy particle collisions.

For this kind of consideration in the t-representation to become realistic, we need a further significant elaboration. We know that there are many different phases in quantum field theory. Consider a system composed of the electron gas in metals, which may have normal and superconducting phases. In the middle of short time events thermal fluctuations may include both of these phases. To deal with this situation the thermal Bogoliubov transformation includes not only tilde and nontilde components (thermal doublets) but also spin-up and spin-down components (Nambu doublets). Then, the thermal Bogoliubov transformation takes the form of four by four matrices [6, 10]. Furthermore, this is not a simple direct product of two sets of two by two matrices, because it should take into account intermediate states between two phases. This illustrates the fact that in general the thermal Bogoliubov transformation takes the form of matrices of high dimension.

As it was pointed out in section 1, the renormalization in TFD means only to calculate many parameters corrected by interaction effects. It is not recommended to choose the corrected parameters for the unperturbed parameters. For example, although the time-space dependent TFD is formulated with time-dependent quantum energy $\omega_k(t)$ for unperturbed energy, a time-independent ω_k for unperturbed energy seems to be more useful. Although the self-energy shift $\delta\omega_k(t)$ depends on time, it simplifies calculations to start with time-independent ω_k and to calculate the time-dependent one as $\omega_k(t) = \omega_k + \delta\omega_k(t)$. We may choose the time-independent unperturbed ω_k to be $\omega_k = \omega_k(t_0)$ with $t_0 = \infty$. In the case of the number density parameter too, one cannot start with the Heisenberg number $n_{Hk}(t)$ for the unperturbed number $n_k(t)$. This is clearly seen from the inspection of equilibrium TFD. In the equilibrium case, the unperturbed n_k should have the Boltzmann distribution, while n_{Hk} is its integration with the spectral function ρ:

$$n_{Hk} = \int d\kappa \, \rho(\kappa) \, n_k(\kappa). \tag{9.160}$$

The function ρ has a peak at $\kappa = \omega_k$ and width κ_k. Since the dissipative constant κ_k does not vanish, ρ is not the delta function. This implies two features. One is that the observed spectrum fluctuates around the equilibrium one with the fluctuation energy of the order of the dissipative constant. Another is that we cannot use n_{Hk} for unperturbed n_k.

In this chapter we considered mostly the TFD for isolated systems. When there is a net flow of dissipative energy to an outside system, it may be appropriate to take the dissipative Hamiltonian as the unperturbed Hamiltonian. In fact, the development of the time-dependent TFD was first performed by referring to the cases with a net flow of dissipative energy to an out-side system [5, 6]. There are applications of TFD for these situations [11, 12, 13, 14, 15].

The introduction of a covariant time derivative suggested a possibility that time-dependent TFD might be extended to a gauge theory. In this book we made a particular choice of parameters α_k and s_k so that we have only one component of the thermal generator $\hat{Q}$. When we choose all three parameters α_k, s_k and n_k dependent on time, we have three generators forming a SU(1,1) algebra. A possibility of SU(1,1) gauge theory and its topological structure was studied by Henning and his collaborators [8]. Even with only one parameter n_k, there might be a possibility of formulating a gauge theory and its associated topology in the $(\vec{x}, t)$-space. These are interesting possibilities.

In this book we tried to build TFD mostly on the basis of concepts of quantum field theory. In parallel to such efforts there have been attempts to cultivate TFD by taking advantages of the knowledge in statistical mechanics. The framework of time-dependent TFD is wide enough to take account of stochastic equations with a random force operator such as the Langevin equation for quantum fields and the stochastic Liouville equation [16, 17]. In addition to the treatment given in this chapter, studies in this direction may reveal a deeper understanding of thermal phenomena, just as Einstein's stochastic interpretation of thermal phenomena gave us a more intuitive understanding of heat than the interpretation within Gibbs' rather technical ensemble theory. There also has been an attempt to introduce into time-dependent TFD some of the concepts of non-equilibrium thermodynamics with emphasis on the hydrodynamical stage [18]. The article [19] presents an excellent description of an attempt to develop a nonequilibrium TFD on the basis of the nonequilibrium statistical mechanics.

When a complete form of TFD is established, we will have a unified view of microscopic and macroscopic worlds including thermal effects.

9.6 REFERENCES

[1] H. Umezawa and Y. Yamanaka, Temporal Description of Thermal Quantum Fields, *Univ. of Alberta Preprint* (1992).

[2] Y. Yamanaka, H. Umezawa, K. Nakamura and T. Arimitsu, Thermo Field Dynamics in t-representation, *Univ. of Alberta Preprint* (1992).

[3] K. Nakamura, H. Umezawa and Y. Yamanaka, Time-space dependent formulation in Thermo Field Dynamics, *Univ. of Alberta Preprint* (1992).

[4] K. Watanabe, K. Nakamura and H. Ezawa, in H. Ezawa, T. Arimitsu and Y. Hashimoto, editors, *Proceedings of the 2nd Workshop on Thermal Field Theories and Their Applications*, (North-Holland, Amsterdam, 1991), p. 79.

[5] T. Arimitsu and H. Umezawa, *Prog. Theor. Phys.* **74** (1985) 429.

[6] T. Arimitsu, M. Guida and H. Umezawa. *Physica* **148A** (1988) 1.

[7] H. Umezawa and Y. Yamanaka, *Advances in Physics* **37** (1988) 531.

[8] P. A. Henning, M. Graf and F. Matthäus, *Physica* **A182** (1992) 489.

[9] St. Mrowczyuski and P. Danielewicz, *Nucl. Phys.* **B342** (1990) 345.

[10] I. Hardman, H. Umezawa and Y. Yamanaka, In K. L. Kowalski, N. P. Landsman and Ch. G. van Weert, editors *Themal Field Theories and Their Applications, Physica* **158A** (1989) 326.

[11] M. Ban and T. Arimitsu, *Physica* **146A** (1987) 89.

[12] T. Tominaga, M. Ban, T. Arimitsu, J. Pradko and H. Umezawa, *Physica* **149A** (1988) 26.

[13] T. Tominaga, T. Arimitsu, J. Pradko and H. Umezawa, *Physica* **150A** (1988) 97.

[14] T. Iwasaki, T. Arimitsu and F. H. Willeboordse. In H. Ezawa, T. Arimitsu, and Y. Hashimoto, editors, *Thermal Field Theories*, (North-Holland, Amsterdam, 1991) p. 469.

[15] T. Arimitsu, F. H. Willeboordse and T. Iwasaki, An Analytical Treatment of a Localized Electron-Phonon System within Non-Equilibrium Thermo Field Dynamics — Intensity Distribution —, *Univ. of Tsukuba Preprint* (1991).

[16] T. Arimitsu, *Phys. Lett.* **153A** (1991) 163.

[17] T. Arimitsu, In H. Ezawa, T. Arimitsu, and Y. Hashimoto, editors, *Themal Field Theories*, (North-Holland, Amsterdam, 1991) p. 207.

[18] T. Arimitsu, Thermal processes in the hydrodynamic stage in terms of non-equilibrium thermo field dynamics. *J. of Phys. A*, in press, 1991.

[19] D. N. Zubarev and M. V. Tokarchuk, *Teoreticheska i Matematicheskaya Fizika* **88** (1991) 286.

Index

Printed in the United States
By Bookmasters